Micelles
Theoretical and Applied Aspects

Micelles
Theoretical and Applied Aspects

Yoshikiyo Moroi

Department of Chemistry
Faculty of Science
Kyushu University
Fukuoka, Japan

PLENUM PRESS • NEW YORK AND LONDON

CHEMISTRY

.4612851

Library of Congress Cataloging-in-Publication Data

Moroi, Yoshikiyo, 1941-
    Micelles : theoretical and applied aspects / Yoshikiyo Moroi.
        p.   cm.
    Includes bibliographical references and index.
    ISBN 0-306-43996-4
    1. Micelles.   I. Title.
QD549.M686   1992
                                                        91-48131
                                                            CIP

ISBN 0-306-43996-4

© 1992 Plenum Press, New York
A Division of Plenum Publishing Corporation
233 Spring Street, New York, N.Y. 10013

Printed in the United States of America

Preface

Almost thirty years ago the author began his studies in colloid chemistry at the laboratory of Professor Ryohei Matuura of Kyushu University. His graduate thesis was on the elimination of radioactive species from aqueous solution by foam fractionation. He has, except for a few years of absence, been at the university ever since, and many students have contributed to his subsequent work on micelle formation and related phenomena. Nearly sixty papers have been published thus far. Recently, in search of a new orientation, he decided to assemble his findings and publish them in book form for review and critique. In addition, his use of the mass action model of micelle has received much criticism, especially since the introduction of the phase separation model. Many recent reports have postulated a role for Laplace pressure in micellization. Although such a hypothesis would provide an easy explanation for micelle formation, it neglects the fact that an interfacial tension exists between two macroscopic phases. The present book cautions against too ready an acceptance of the phase separation model of micelle formation.

Most references cited in this book are studies introduced in small group meetings of colloid chemists, the participants at which included Professors M. Saito, M. Manabe, S. Kaneshina, S. Miyagishi, A. Yamauchi, H. Akisada, H. Matuo, M. Sakai, and Drs. O. Shibata, N. Nishikido, and Y. Murata, to whom the author wishes to express his gratitude for useful discussions. The author also was in residence at the University of Wisconsin–Madison for sixteen months; his collaboration with Professor P. Mukerjee and Mr. M. Murata as a research associate greatly aided in the development of the concepts described in Chapters 3 and 4. His study for fifteen months at L'École Polytechnique Fédéral de Lausanne with Professor M. Grätzel and Drs. P. P. Infelta, R. Humphry-Baker, and A. Braun gave rise to Chapter 12. Chapters 5, 8, and 13 owe a great deal to Professors M. Tanaka; K. Motomura; and I. Satake and K. Hiramatsu, respectively. Collaboration with Professor T. Kuwamura and Dr. S. Inokuma contributed greatly to the concept of micelle temperature range as discussed in Chapter 6. It is a pleasure for the author to acknowledge his indebtedness to Professor J. Kratohvil of Clarkson University, who read the manuscript and offered criticism and advice. The author is also grateful to Miss Y. Kamishiro for her helpful assistance. This book is thus a mixed crystal of collaborations and contributions from all the persons mentioned above. The author also expresses his thanks to those predecessors whose work has made colloid science so fascinating. In addition, any criticism of readers on the contents of this book is welcomed.

Finally, the author wishes to thank his parents, for giving him the opportunity for advanced study, and his wife, Noriko, for her patience and encouragement during the preparation of this book.

Yoshikiyo Moroi

Fukuoka, Japan

Contents

1

Introduction

Forty years have passed since the publication of three famous books bearing the same title—*Colloid Science*. The first was by A. E. Alexander and P. Johnson,[1] the second by J. W. McBain,[2] and the third by H. R. Kruyt.[3] Since then, colloid science has generated a number of new but related fields, such as studies of polymers, semiconductors, liquid crystals, membranes, and vesicles. Micellar aggregates have served as an important bridge between microscopic and macroscopic chemical species in the development of new technologies. The importance of colloid science is now fully recognized, and it has become established as the basic foundation of nearly all fields of solution science.

The thermodynamic equilibria of amphiphilic molecules in solution involve four fundamental processes: (1) dissolution of amphiphiles into solution; (2) aggregation of dissolved amphiphiles; (3) adsorption of dissolved amphiphiles at an interface; and (4) spreading of amphiphiles from their bulk phase directly to the interface (Fig. 1.1). All but the last of these processes are presented and discussed throughout this book from the thermodynamic standpoint (especially from that of Gibbs's phase rule), and the type of thermodynamic treatment that should be adopted for each is clarified. These discussions are conducted from a theoretical point of view centered on dilute aqueous solutions; the solutions dealt with are mostly those of the ionic surfactants with which the author's studies have been concerned. The theoretical treatment of ionic surfactants can easily be adapted to nonionic surfactants. The author has also concentrated on recent applications of micelles, such as solubilization into micelles, mixed micelle formation, micellar catalysis, the protochemical mechanisms of the micellar systems, and the interaction between amphiphiles and polymers. Fortunately, almost all of these subjects have been his primary research interests, and therefore this book covers, in many respects, the fundamental treatment of colloidal systems.

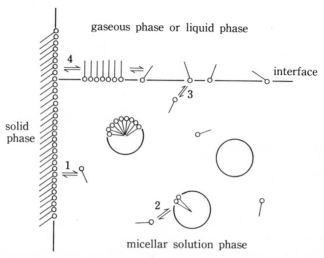

gaseous phase or liquid phase

Figure 1.1. Four fundamental processes for thermodynamical equilibria of amphiphilic molecules in solutions: (1) dissolution of amphiphiles into solution, (2) aggregation of dissolved amphiphiles, (3) adsorption of dissolved amphiphiles at an interface, (4) spreading of amphiphiles from their own bulk phase directly to an interface.

The number of scientific reports relating to amphiphilic molecules has greatly expanded during the past 20 years. Amphiphilic molecules, or *amphiphiles*, are not only highly interesting from the physicochemical viewpoint but also are fundamental to life: it is no exaggeration to say that living things are made up of colloids comprising a wide variety of amphiphiles. In this monograph, however, attention focuses on the aggregation of amphiphiles in solution. Many kinds of amphiphiles form molecular aggregates in solution above a narrow concentration range. These aggregates are called *micelles*, and the concentration range above which they form is called the *critical micelle concentration* (CMC).

Many monographs,[4-8] proceedings,[9-11] review articles,[12-15] and scientific papers have been published on micelle formation. It has been considered from two primary viewpoints: one regarding the micelle as a chemical species (the *mass action model*), and the other considering it as a separate phase (the *phase separation model*). The two models converge as the aggregation number of micelles under observation increases; the problem lies in determining which model is more appropriate for micelles whose aggregation number is less than a few hundred.

Many recent reports on micelle formation and solubilization into micelles have treated micelles as a separate phase.[16-20] With regard to solubilization, in particular, an increased pressure within the micelle, in accordance with Laplace's law, has often been postulated to explain a diminished transfer of free energy per methylene group from the aqueous medium into the micelle interior compared with the free energy transfer from an aqueous medium into bulk liquid hydrocarbon.[16,17] The decreased free energy transfer per methylene group associated with micelle formation has also been attributed to partial crystallization of the alkyl chain in the interior of the micelle, caused by the same Laplace-induced pressure increase.[21] An interfacial tension exists at the boundary between two bulk phases.[22-24] Therefore, a pressure increase can occur only when micelles are a separate phase. If interfacial tension were present within the micelle, a difference would be observed in the association constant ($\bar{K}_1$) between a solubilizate (R) and a vacant micelle (M), depending on whether the solubilization site lay inside or outside a plane of interfacial tension (see Chapter 9):

$$M + R \xrightleftharpoons{\bar{K}_1} MR_1 \qquad (1.1)$$

This difference might be seen by using solubilizate molecules with hydrophobic akyl chains of varying lengths as probes, since the alkyl chain is believed to be located partly inside and partly outside the micellar plane of tension. Figure 1.2 shows the change in the solubilization constant observed in 4-n-alkylbenzoic acid and dodecylsulfonic acid micelles when alkyl chains of varying length were used.[22] The alkyl chains of the solubilizates ranged in length from C_0 to C_8, where the C_8 chain is almost as long as the surfactant molecule. Therefore, if a plane of tension were located inside the micelle, the alkyl chains long enough to penetrate the micelle core would encounter an increased Laplace pressure, and the plot in Fig. 1.2 should become less steep above a certain carbon number. The experimental data did not show this effect, indicating that there probably is not a plane of interfacial tension in the micelle.

Discussions of micelle formation and related phenomena based on the phase rule also lead to important conclusions, because this approach is thermodynamically correct. The concept of degrees of freedom, which is based on this approach, is employed frequently in this book.

Colloid chemistry is used in many fields of science and technology, as is evidenced both by the voluminous literature on the subject and by the proliferation of research institutions. The audience for this book therefore includes senior undergraduate and graduate students in chemistry, applied chemistry, and pharmacology as well as researchers in surfactants and detergents. The author hopes that this book will provide preliminary guidance to the many interrelated fields centered on micelle chemistry, and that it will be useful to workers in the fields of biology, medicine, pharmacology, and of course, colloid chemistry.

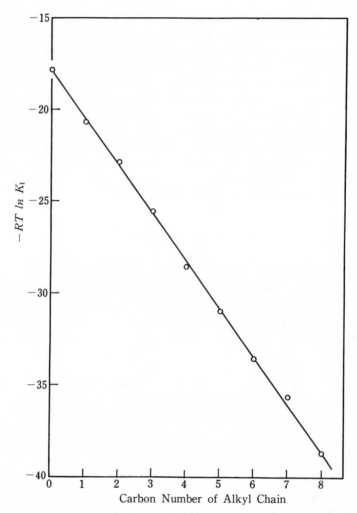

Figure 1.2. Standard free energy change for the association constant between solubilizate and micelles plotted against carbon number of the solubilizate alkyl chain. The solubilizate is 4-n-alkylbenzoic acid and the micelle is made of dodecylsulfonic acid.[22] (Reproduced with permission of Academic Press.)

Chapters 1 through 8 deal with fundamental theories of amphiphilic molecules, with a focus on micelles, and Chapters 9 through 13 cover the applications of micelles. All of the chapters, however, cover the necessary fundamental concepts—including relevant symbols and equations—and are self-contained. Concentrations are ususally expressed by the italic letter C

with a subscript indicating the chemical species, but the square-bracket convention [ ] is also used for complicated species.

References

1. A. E. Alexander and P. Johnson, *Colloid Science*, Oxford University Press (Clarendon), London (1949).
2. J. W. McBain, *Colloid Science*, Heath, Boston (1950).
3. *Colloid Science* (H. R. Kruyt, ed.), Elsevier, Amsterdam (1952).
4. K. Shinoda, T. Nakagawa, B. Tamamushi, and T. Isemura, *Colloidal Surfactants*, Academic Press, New York, 1(63).
5. C. Tanford, *The Hydrophobic Effect: Formation of Micelles and Biological Membranes*, Wiley, New York, 1(973).
6. *Aggregation Processes in Solution* (E. Wyn-Jones and J. Gormally, eds.), Elsevier, Amsterdam (1983).
7. *Nonionic Surfactants* (M. J. Schick, ed.), Dekker, New York (1967).
8. *Cationic Surfactants* (E. Jungermann, ed.), Dekker, New York (1970).
9. International Congress of Surface Activity: 1st proceedings, Paris (1954); 2nd proceedings, Butterworths, London (1957); 3rd proceedings, Cologne (1960); 4th proceedings, Brussels (1964); 5th proceedings, Barcelona (1968); 6th proceedings, Zurich (1972); 7th proceedings, Moscow (1976).
10. International Symposium on Surfactants in Solution: 1. *Micellization, Solubilization, and Microemulsions* (K. L. Mittal, ed.), Plenum Press, New York (1977), 2. *Solution Chemistry of Surfactants* (K. L. Mittal, ed.), Plenum Press, New York (1979), 3. *Solution Behavior of Surfactants: Theoretical and Applied Aspects* (K. L. Mittal and E. J. Fendler, eds.), Plenum Press, New York (1982), 4. *Surfactants in Solution* (K. L. Mittal and B. Lindmann, eds.), Plenum Press, New York (1984), 5. *Surfactants in Solution* (K. L. Mittal and P. Bothorel, eds.), Plenum Press, New York (1986).
11. *Physics of Amphiphiles: Micelles, Vesicles and Microemulsions, Proceedings of the International School of Physics "Enrico Fermi"* (V. Degiorgio and M. Corti, eds.), North-Holland, Amsterdam (1985).
12. P. Murkerjee, *Adv. Colloid Interface Sci. 1*, 241 (1967).
13. G. C. Kresheck, in: *Water: A Comprehensive Treatise*, Vol. 4, *Aqueous Solutions of Amphiles and Macromolecules* (F. Franks, ed.), p. 95, Plenum Press, New York (1975).
14. H. Wennerstrom and B. Lindman, *Phys. Rep. 52*, 1 (1979).
15. L. R. Pratt, B. Owenson, and Z. Sun, *Adv. Colloid Interface Sci. 26*, 69 (1986).
16. I.B.C. Matheson and A. D. King, Jr., *J. Colloid Interface Sci. 66*, 464 (1978).
17. W. Prapaitrakul and A. D. King, Jr., *J. Colloid Interface Sci. 106*, 186 (1985).
18. E. Ruckenstein and R. Krishnan, *J. Colloid Interface Sci. 71*, 321 (1979).
19. E. Lissi, E. Abuin, and Z. N. Rocha, *J. Phys. Chem. 84*, 2406 (1980).
20. J. C. Ericksson, S. Ljunggren, and U. Henriksson, *J. Chem. Soc. Faraday Trans. 2, 81*, 833 (1985).
21. P. Murkerjee, *Kolloid Z. Z. Polym. 236*, 76 (1970).
22. Y. Moroi and R. Matuura, *J. Colloid Interface Sci. 125*, 463 (1988).
23. R. Defay, I. Prigogine, A. Bellemans, and D. H. Everett, *Surface Tension and Adsorption*, Longmans, London (1966).
24. E. A. Guggenheim, *Thermodynamics: An Advanced Treatment for Chemists and Physicists*, North-Holland, Amsterdam (1977).

2

Surface-Active Agents

2.1. Classification

Surface-active agents, or *surfactants*, owe their name to their interesting behavior at surfaces and interfaces. They are positively adsorbed at interfaces between phases, and the adsorption of surfactant lowers the interfacial tension between the phases (see Chapter 8). Because of their ability to lower interfacial tension, surfactants are used as emulsifiers, detergents, dispersing agents, foaming agents, wetting agents, penetrating agents, and so forth.

Many types of substances act as surfactants, but all share the property of *amphipathy*: the molecule is composed of a nonpolar hydrophobic portion and a polar hydrophilic portion,[1] and is therefore partly hydrophilic and partly hydrophobic. Surfactants may be referred to as either *amphiphilic* or *amphipathic*; the terms are synonymous. The polar, hydrophilic part of the molecule is called the *hydrophilic* or *lipophobic* group, and the nonpolar, hydrophobic part is called the *hydrophobic* or *lipophilic* group. Often the hydrophilic part of the molecule is simply called the *head* and the hydrophobic part—usually including an elongated alkyl substituent—is called the *tail*. The presence of a hydrophilic group makes surfactants slightly soluble in aqueous media, and is central to the physicochemical properties of aqueous surfactant solutions.

Surfactants are classified on the basis of the charge carried by the polar head group as *anionic, cationic, nonionic*, or *amphoteric*. Tables 2.1 through 2.4 show the chemical structures of typical examples of these classes. *Lecithin, cephalin*, and the *bile acids* are ususally classified as *biosurfactants*. The bile acids and their conjugates have different properties in solution from surfactants with a long alkyl chain.[28-30]

Table 2.1. Chemical Structure of Hydrophilic Groups for Anionic Amphiphiles

Chemical structure[a]	Name
$R-(COO^-)_n M^{n+}$	Carboxylate
$R-COO^- M^{2+}$ $\|$ SO_3^-	Sulfocarboxylate
$R-COO^- M^{(n+1)+}$ $\|$ OPO_3H_{3-n}	Phosphonocarboxylate
$R-CON(CH_3)CH_2COO^- M^+$	Sarcoside
$R-OSO_3^- M^+$	Sulfate
$R-(OCH_2CH_2)_n - OSO_3^- M^+$	Polyoxyethylene sulfate[2,3]
$R-SO_3^- M^+$	Sulfonate
$R-(OCH_2CH_2)_n-SO_3^- M^+$	Polyoxyethylene sulfonate
$R-CH-SO_3^- M^+$ $\|$ CH_2OH	1-Hydroxy-2-sulfonate
R—⟨◯⟩—SO_3^- M^+	Benzene sulfonate[4,5]
R—⟨◯◯⟩—SO_3^- M^+	Naphthalene sulfonate
$R-OPO_3H_{3-n} M^{n+}$	Phosphate[6,7]

[a] R—: long hydrophobic tail.

2.2. Hydrophile–Lipophile Balance

The term *hydrophile-lipophile balance* (HLB), first suggested by Clayton, refers to the balance in size and strength between the hydrophilic and hydrophobic parts of a surfactant molecule.[31] Griffin later developed the concept of the HLB for emulsifiers on the basis of their aqueous solubility.[32,33] The *HLB value* is an empirical number assigned to nonionic surfactants on the basis of a wide variety of emulsion experiments carried out on surfactants at the Atlas Powder Company. These HLB values range from 1 to 40, the low numbers generally indicating solubility in oil and the high numbers solubility in water. Nevertheless, emulsifiers with the same HLB value may differ in solubility.

An emulsifier has two different actions: it promotes the formation of an emulsion, and it determine whether an *oil/water* (O/W) or a *water/oil* (W/O) emulsion will be formed. The second action is closely connected

Table 2.2. Chemical Structure of Hydrophilic Groups for Cationic Amphiphiles

Chemical structure[a]	Name
$$R-\overset{\displaystyle R_1}{\underset{\displaystyle R_3}{\overset{|}{\underset{|}{N^+}}}}-R_2 \; X^-$$	Ammonium[8]
$$R-\overset{+}{S}-R_1 \; X^- \\ \quad \underset{|}{\;} \\ \quad R_2$$	Sulfonium
$$R-\overset{\displaystyle R_1}{\underset{\displaystyle R_3}{\overset{|}{\underset{|}{P^+}}}}-R_2 \; X^-$$	Phosphonium
	Pyridinium[9]
	Quinolinium
	Viologen[10]

[a]R—: long hydrophobic tail; R_1—, R_2—, R_3—: hydrogen or short alkyl chain.

with the HLB value. On the basis of systematic emulsion experiments, Griffin found that the HLB values of mixtures of two or more emulsifiers are additive[32]: the HLB value of the mixture is equal to the sum of the HLB values of the constituents multiplied by their weight fractions in the mixture x_i^w:

$$HLB = \sum_i x_i^w (HLB)_i \qquad (2.1)$$

Griffin also listed several estimated HLB values for emulsifiers, which had been determined and correlated by an extensive series of emulsifier blending tests (Table 2.5). Using these values and Eq. (2.1), it is possible to determine an HLB value for any surfactant by blending it with a surfactant of known HLB value. In addition, the following simple formulas for calculating HLB values were derived from systematic tests.[33]

For most polyhydric alcohol fatty acid esters (sorbitan monoester type), approximate HLB values may be obtained using the following equation:

$$HLB = 20(1 - S/A) \qquad (2.2)$$

Table 2.3. Chemical Structure of Hydrophilic Groups for Nonionic Amphiphiles

Chemical structure[a]	Name
$R-(OCH_2CH_2)_n-OH$	Polyoxyethylene alcohol[11]
$R-(OCH_2CH_2CH_2)_n-OH$	Polyoxypropylene alcohol[12]
$R-COO-(CH_2CH_2O)_n-H$	Polyoxyethylene ester
$R-COO-\underset{\underset{OH}{\mid}}{CH}-CH_2OH$	Glycerol monoester
$R-COO-CH_2\underset{\underset{CH_2OH}{\mid}}{\overset{\overset{CH_2OH}{\mid}}{C}}-CH_2OH$	Pentaerythritol monoester
$R-COO-CH_2-CH\overset{O}{\diagdown}CH_2$ with $HO-CH$, $CH-OH$, $CH-OH$	Sorbitan monoester
$R-(CH_2CH_2O)_n$	Crown ether[13,14]
$R-\underset{\downarrow}{S}-R_1$, O	Sulfoxide[15,16]
$R-\underset{\downarrow}{S}-(CH_2)_n-OH$, O	Sulfinyl alkanol[15,17]
$R-S-(CH_2CH_2O)_n-H$	Polyoxyethylene thioether
$R-\underset{\underset{R_2}{\mid}}{\overset{\overset{R_1}{\mid}}{N}}\rightarrow O$	Amine oxide[18-21]
$R-(CH_2CH_2NH)_n$	Azacrown[22]
$R-\underset{\underset{R_2}{\mid}}{\overset{\overset{R_1}{\mid}}{P}}\rightarrow O$	Phosphine oxide[23,24]
$R-CO-\underset{}{N}-CH_2\overset{CH_3}{CH}-CH-CH-CH-CH_2OH$ with OH, OH	N-Methylglucamine[25-27]

[a]R—: long hydrophobic tail; R_1-, R_2-: hydrogen or short alkyl chain.

Table 2.4. Chemical Structure of Hydrophilic Groups for Amphoteric Amphiphiles

Chemical structure[a]	Name
$R-CH-^+N-R_2$ with R_1 above N, R_3 below N, and COO^- below CH	C betaine
$R-^+N-CH_2COO^-$ with R_1 above and R_2 below	N betaine
$R-^+N-CH_2COOH$ with CH_2COOH above and CH_2COO^- below	Triglycine
$R-^+N-CH_2CH_2SO_3^-$ with R_1 above and R_2 below	N, N-Dialkyl taurine
$R-\cdots-O-P-O-CH_2CH_2-^+N-CH_3$ with O^- above P, O below P, CH_3 above N, and CH_3 below N	Phosphatidylcholine[4]

[a]R—: long hydrophobic tail; R_1—, R_2—, R_3—: hydrogen or short alkyl chain.

where S is the saponification number of the ester and A is the acid number of the acid. This equation can be written as

$$HLB = 20(1 - M_h/M_w) \qquad (2.2')$$

where M_h is the weight of the hydrophobic group and M_w is the molecular weight.[34]

Many fatty acid esters (Tween type) do not give good saponification number data. For these substances, the HLB values may be calculated by

$$HLB = (E + P)/5 \qquad (2.3)$$

where E is the weight percent of oxyethylene and P is the weight percent of polyhydric alcohol. If the hydrophilic group consists only of polyoxyethylene (Igepal type), Eq. (2.3) simplifies to

$$HLB = E/5 \qquad (2.3')$$

As mentioned above, Griffin proposed an HLB scale for emulsifiers ranging from 1 (very lipophilic) to 40 (very hydrophilic). HLB values were assigned by determining the proportions of different emulsifier combinations

Table 2.5. Estimated HLB Values for Various
Types of Emulsifiers[a]

Emulsifier		Estimated HLB
	Anionic	
TEA oleate		12
Sodium oleate		18
Potassium oleate		20
	Cationic	
Atlas G-251		25–35
	Nonionic	
Oleic acid		~1
Span 85		1.8
Span 80		4.3
Span 60		4.7
Span 20		8.6
Tween 81		10.0
Tween 60		14.9
Tween 80		15.0
Tween 20		16.7

[a]Reproduced with permission of the Society of Cosmetic
Chemists.[32]

needed to make the best oil/water emulsions; 75 emulsions were used to determine the HLB value of each surfactant. Atlas Chemical Industries (now ICI America) has since recommended that an initial series of nine test emulsions be prepared to yield an approximate HLB value, which is then refined with further emulsions.[35] This arduous method is applicable only to nonionic surfactants. HLB values obtained according to this protocol range from 1 (most lipophilic) to 20 (most hydrophilic). Suitable applications and HLB ranges are shown in Table 2.6.[36]

In another attempt to overcome the limitations of Griffin's procedure, Davies[37] attempted to calculate HLB values by assigning an *HLB contribution group number* to each functional group in a molecule after studying the relative coalescence rates of stabilized oil droplets in water and water droplets in oil. Table 2.7 gives the group numbers characteristic of each functional group. Davies's equation, which is applicable to ionic as well as nonionic surfactants, is

$$\text{HLB} = \sum (\text{hydrophilic group number})$$
$$- \sum (\text{hydrophobic group number}) + 7 \qquad (2.4)$$

Davies's method is useful if the structure and proportions of the components in the surfactant are known. The greatest disadvantage of the method arises

Table 2.6. HLB Ranges and Suitable Applications[a]

HLB range	Applications
1 to 6	Water-in-oil emulsifier
6 to 9	Wetting agent
8 to 18	Oil-in-water emulsifier
13 to 15	Detergent
15 to 18	Solubilizer
	HLB by dispersibility
1 to 4	No dispersibility in water
3 to 6	Poor dispersion
6 to 8	Milky dispersion after vigorous agitation
8 to 10	Stable milky dispersion
10 to 13	Translucent to clear dispersion
>13	Clear solution

[a]Reproduced with permission of Dekker.[36]

from the fact that the contribution of a given hydrophilic group to the polarity of a surfactant molecule tends to decrease as the size of the molecule increases. The method was developed further by Lin *et al.*[34,38]

Later, Greenwald *et al.* developed a classification system based on liquid–liquid distribution coefficients of surfactants in water and isooctane.[39] In 1962, Huebner introduced a quantity called the *polarity index* (PI) intended to replace the HLB value.[40] This index was found to have a linear relationship with the HLB value.[41] The polarity index is determined from the carbon number corresponding to methanol, when methanol and normal hydrocarbons are separated on a gas chromatograph with the surfactant as

Table 2.7. HLB Group Numbers for Hydrophilic and Hydrophobic Groups[a]

Hydrophilic	Group number	Hydrophobic	Group number
$-SO_4Na$	38.7	$-CH-$	0.475
$-COOK$	21.1	$-CH_2-$	0.475
$-COONa$	19.1	$-CH_3$	0.475
$-SO_3Na$	11.0	$=CH-$	0.475
N (tertiary amine)	9.4	$-CF_2-$	0.870
Ester (free)	2.4		
$-COOH$	2.1		
$-OH$ (free)	1.9		
$-O-$	1.3		
$-OH$ (sorbitan ring)	0.5		

[a]Reproduced with permission of Academic Press.[34]

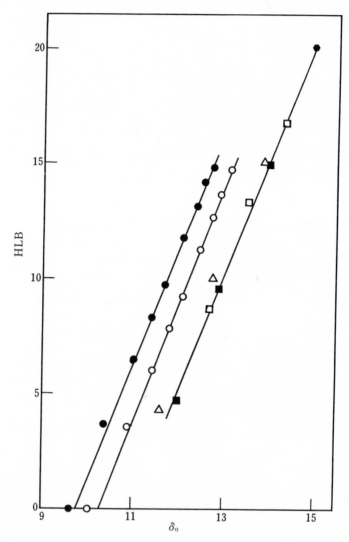

Figure 2.1. Relationship between HLB and δ_0 for three classes of polyoxyethylated surfactants: (●) dodecanol derivatives; (○) octylphenol derivatives; (△) sorbitan monooleate derivatives; (■) sorbitan monostearate derivatives; (□) sorbitan monolaurate derivatives; (◆) polyethylene glycol 3350.[46] (Reproduced with permission of the American Pharmaceutical Association.)

a stationary phase. Huebner's formula for the polarity index is

$$PI = 100 \log(n_c - 4.7) + 60 \qquad (2.5)$$

where n_c is the apparent number of carbon atoms in a standard alkane having the same retention time as methanol (obtained from a graph of the hydrocarbon retention times plotted against the number of carbon atoms in the hydrocarbons); 4.7 is a statistically determined factor; and 60 is the value needed to make the index positive. Polarity index values for surfactant mixtures are additive in the same way as HLB values [see Eq. (2.1)].

The polarity index was further investigated by Becher and Birkmeier, who defined the polarity of a surfactant in an inverse fashion as *the ratio of the retention times of a polar and a nonpolar material.*[42] Gas chromatographic investigations on the polarity of surfactants are still being carried out.[43,44]

Small introduced the *solubility parameter*, an additive constitutive property of a surfactant molecule that can be calculated from the additive contribution of its functional groups.[45] Schott developed this concept and introduced the overall solubility parameter δ_0, made up of three components[46]:

$$\delta_0 = (\delta_D^2 + \delta_P^2 + \delta_H^2)^{1/2} \qquad (2.6)$$

where δ_D reflects dispersion forces, σ_P reflects dipole–dipole forces, and δ_H reflects hydrogen bonding forces. Overall, solubility parameters show a good linear relationship with HLB values (Fig. 2.1). Surprisingly, the curves for the three structurally dissimilar surfactants shown in Fig. 2.1 are spaced only about 1.2 $(\mathrm{cal/cm^3})^{1/2}$ apart. This finding indicates that the nature of the hydrocarbon portion of a nonionic surfactant has only a limited effect on its solubility parameter. This finding also indicates that the objection leveled against HLB values calculated using Eq. (2.3)—that they treat the hydrocarbon portion of the molecule only in terms of its weight percent—is of limited validity.

Many attempts have been made to relate HLB value to the properties of surfacts,[38,47-50] but a definite relation has not been obtained.

2.3. Purification of Surfactants

Impurities are small amounts of foreign chemicals coexisting with the chemical of interest. In practice, impurities often confer desirable properties on a surfactant and are added as regular constituents. For purpose of basic surfactant research, however, all chemical species other than the surfactants of interest must be considered impurities. For example, surfactant molecules homologous to the species under study but differing from it in alkyl chain length constitute impurities. For this reason the hydrophobic reactants used to prepare a pure surfactant sample should themselves be as pure as possible. Other ubiquitous impurities include residual reactants, inorganic chemicals, solvents, and by-products. The effects of these impurities on the physicochemical properties of surfactants are very instructive.

2.3.1. Effects of Impurities

One type of impurity that strongly influences the solution properties of surfactants consists of residual hydrophobic reactants having a polar head group such as $-COOH$, $-OH$, $-NH_2$, $-X$ (where X is a halogen atom), or $-C=CH_2$. Compounds of this type are only sparingly soluble in water, and therefore deserve attention in studies on interface adsorption and micelle formation by surfactants. These compounds are very surface-active and can significantly reduce interfacial tension; moreover, they are easily solubilized into surfactant aggregates or micelles. For example, if sodium dodecyl sulfate (SDS) containing a residual amount of the hydrophobic reactant dodecanol is added gradually to an aqueous phase, it is found that with increasing surfactant, the surface tension of the sample first drops to a minimum and then rises toward the value characteristic of pure SDS. Samples prepared using pure SDS do not show this minimum. The surfactant concentration at the minimum corresponds to the critical micelle concentration (CMC) of the unpurified surfactant, which is lower than the CMC of purified surfactant. The increase in surface tension of the unpurified surfactant above the CMC is caused by the solubilization of dodecanol into SDS micelles and the consequent desorption of dodecanol from the sample surface (Fig. 2.2).[51]

If a hydrophobic reactant includes homologs with different alkyl chain lengths, then the chain length of the product will also be heterogeneous. The CMC values of homologous surfactants are given by[52]

$$\log CMC = B - Dn_c \tag{2.7}$$

where n_c is the number of carbon atoms in the alkyl chain and B and D are the constants of each homolog. Figure 2.3 shows the mixed CMC for the system SDS/sodium tetradecyl sulfate (see Chapter 10 for discussion of mixed micelle formation).[53] As shown by Fig. 2.3, the CMC is increased by the homolog with the shorter alkyl chain and decreased by the homolog

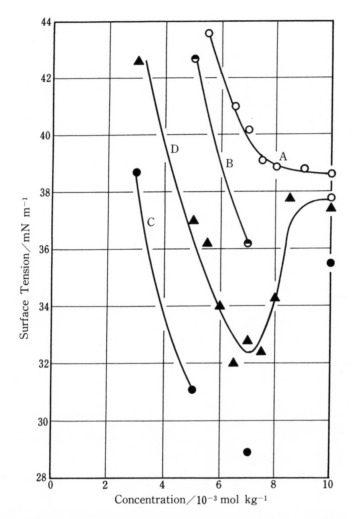

Figure 2.2. Effect of dodecanol (DOH) on the surface tension of solutions of sodium dodecyl sulfate (SDS): A, pure SDS; B, pure SDS + 0.1% DOH of the amount of SDS; C, pure SDS + 0.5% DOH of the amount of SDS; D, SDS before final 36-h Soxhlet extraction with ethyl ether.[51] (Reproduced with permission of the American Chemical Society.)

with the longer alkyl chain, but the effect of the long-chain homolog is much stronger. Long-chain homolog impurities should therefore be eliminated with special care.

Another important type of impurity is inorganic by-products that increase the ionic strength of the surfactant solution. Inorganic impurities

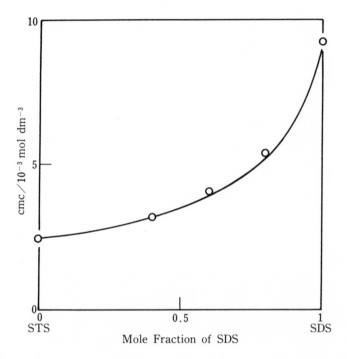

Figure 2.3. CMC change of sodium dodecyl sulfate (SDS)–sodium tetradecyl sulfate (STS) mixture at 47°C: (○), observed value; (—) theoretical curve.[53] (Reproduced with permission of Academic Press.)

of this type strongly influence the interfacial adsorption and micelle formation of ionic surfactants. The decrease in the CMC caused by a specific concentration of a counterion is given by[54]

$$\log \text{CMC} = B' - D' \log C_i \tag{2.8}$$

where C_i is the counterion concentration and B' and D' are constants characteristic of each surfactant. As shown in Fig. 2.4, log CMC decreases linearly with $\log C_i$. Salts have a weaker effect on nonionic surfactants than on ionic surfactants, but still cause a slight decrease in the CMC.

2.3.2. Techniques for Purifying Surfactants

In many cases, a surfactant sample is found to exhibit particular phenomena only over specific concentration ranges. This type of concentration dependence generally reflects the presence of impurities, which can have both a qualitative and a quantitative effect on the behavior of the

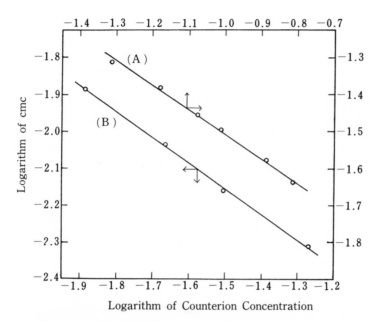

Figure 2.4. Plots of logarithm of CMC against logarithm of counterion concentration: (A) sodium 4-*n*-hexylbenzoate at 25°C; (B) sodium 4-*n*-octylbenzoate at 35°C.

surfactant. Purity is therefore often a crucial factor in surfactant research. This section discusses methods for purifying surfactants, using SDS as a representative ionic surfactant. The purification methods used for SDS can be applied to other ionic surfactants.

Purification techniques handle organic and inorganic impurities separately. In principle, organic impurities can be eliminated by repeated recrystallization from organic solvents, and inorganic impurities can be removed by recrystallization from water. Recrystallization from water is only effective when the micelle temperature range (MTR or Krafft point; see Chapter 6) of the surfactant is above 5°C. Inorganic impurities can alternatively be removed as an insoluble residue by recrystallization from dry organic solvents (e.g., isopropyl alcohol[55]).

Dodecanol (a reactant used in the formation of SDS) is very difficult to remove because it forms a complex with SDS.[56-58] This complex completely decomposes above 66°C. Even continuous extraction by ethyl ether for more than 40 hr using the Soxhlet apparatus does not completely remove dodecanol from this complex; an endothermic peak due to the alcohol is still observable. Foam fractionation, on the other hand, is very effective for removing small amounts of surface-active impurities, including dodecanol.[30,59] Foam fractionation is most effective when the surfactant concentration is below the CMC, because above the CMC surface-active impurities are solubilized into micelles. Foam fractionation is therefore not suitable for purifying large amounts of surfactant. Another effective method for removing dodecanol is evacuation under reduced pressure in an inactive gas atmosphere above 66°C.[56,60] However, this method is slow and permits some thermal decomposition of the surfactant, so it should be performed as a last step to eliminate a small amount of residual alcohol.

Another method for removing dodecanol is alternating batch extraction and recrystallization from water. The extraction steps should be performed for more than 8 hr using a completely dried sample to maximize contact between the ether and the surfactant. The ether extractions remove the alcohol near the crystal surface, whereas the recrystallization from water not only removes inorganic impurities but also dilutes the dodecanol concentrated within the crystals. Three iterations of this sequence followed by heat treatment is ususally sufficient to eliminate the minimum from the curve of surface tension versus concentration. This method is applicable to all ionic surfactants.

The hydrophilic portions of nonionic surfactants ususally consist of a polyoxyethylene group formed by successive addition of ehylene oxide to a functional group on the long hydrophobic chain. To purify this type of nonionic surfactant, it is necessary to know the size distribution of the polyoxyethylene groups. Flory found that these groups display a Poisson distribution (see below).[61]

Nonionic surfactants of the polyoxyethylene type are typically produced by reaction of ethylene oxide with an alcohol:

$$R—OH + CH_2—CH_2 \rightarrow R—O—CH_2CH_2O—H\cdots$$
$$\diagdown \diagup$$
$$O$$

$$\rightarrow R—O—(CH_2CH_2O)_n—H \qquad (2.9)$$

The molecular size distribution of the polar head depends on the number of oxyethylene units that add to the parent alcohol during polymerization. The polymerization process should fulfill two conditions: (1) the total number N_t of molecules that have propagating functional groups should remain constant throughout the reaction, and (2) the chains must be constructed by a series of kinetically identical monomer addition reactions. Let $N_0, N_1, N_2, \ldots$ be the numbers of species $m_0, m_1, m_2, \ldots$ in a specified reaction volume that has zero, one, two, ... added oxyethylene units. The time dependence of N_0 is then expressed by

$$dN_0/dt = -fN_0 \tag{2.10}$$

where f is some function that depends on the kinetics of the process and the manner in which it is carried out. Presumably, f, which includes the velocity constant, will be a function of the ethylene oxide concentration, but it may also be a funtion of other variables such as time and the number of propagating molecules N_t ($= \sum N_i$). Likewise, the following equations apply to N_1 and N_i:

$$dN_1/dt = fN_0 - fN_1 \tag{2.11}$$

$$dN_i/dt = fN_{i-1} - fN_i \tag{2.12}$$

On the other hand, the rate of disappearance of ethylene oxide is

$$-dm/dt = fN_t \tag{2.13}$$

where m is the number of ethylene oxide monomers in the volume. The ratio at time t of the number of ethylene oxide monomers consumed to N_t, or the average number ν of oxyethylene units per surfactant molecule, is given by

$$\nu = \Delta m/N_t = \int_0^t f \, dt \tag{2.14}$$

or

$$d\nu = f \, dt \tag{2.14'}$$

Substituting (2.14') into (2.10) and (2.12), we obtain

$$dN_0 = -N_0 \, d\nu \tag{2.15}$$

$$dN_i = (N_{i-1} - N_i) \, d\nu \tag{2.16}$$

Integration of (2.15) yields

$$N_0 = N_t \exp(-\nu) \tag{2.17}$$

Substituting (2.17) into (2.16) with $i = 1$ and integrating the resulting differential equation, we obtain

$$N_1 = N_t \, \nu \exp(-\nu) \tag{2.18}$$

Likewise,

$$N_2 = N_t(\nu^2/2) \exp(-\nu) \tag{2.19}$$

Continuation of the process yields

$$N_i / N_t = \nu^i \exp(-\nu)/i! \tag{2.20}$$

This general expression for the mole fraction of the i-mer in a surfactant sample follows the Poisson distribution.

Purification of surfactants from a reaction mixture of this type involves either isolating the surfactant with a monodisperse polyoxyethylene group or lowering the molecular weight distribution. The by-product polyoxyethylene glycol can be removed by extraction from the reaction mixture in a 5 M NaCl solution into ethyl acetate[62] or n-butanol,[63] and nonionic surfactants of higher molecular weight can be eliminated by ultrafiltration.[64] Narrowing the molecular weight distribution by molecular distillation is feasible only for species of relatively low molecular weight,[65-67] because long polyoxyethylene chains may decompose under the distillation conditions.[64] The alkyl polyoxyethylene monoethers can be purified by vacuum distillation[65-67] or chromatographic separation,[65,68] provided the number of oxyethylene unit is relatively small. These compounds are particularly prone to oxidation.[65] In principle, nonionic surfactants with different oxyethylene units can be separated by chromatography using silicic columns with chloroform–acetone eluents mixed in graded ratios.[69] However, a wide variety of nonionic surfactants with homogeneous polyoxyethylene units are available commercially from Nikko Chemicals Inc. (Japan). A series of booklets that assemble the publications on these nonionic surfactants has been published: the first covers fundamental research, for example on physicochemical and surface chemical properties; the second covers applied research, for example emulsification, solubilization, dispersion, and wetting; and the third coves chemical and biochemical reactions.[70]

References

1. P. Mukerjee and K. J. Mysels, *Natl. Stand. Ref. Data Ser.* (U.S. Natl. Bur. Stand.) No. 36 (1971).
2. D. Attwood and A. T. Florence, *Kolloid Z. Z. Polym. 246*, 580 (1971).
3. M. Hato and K. Shinoda, *J. Phys. Chem. 77*, 378 (1973).
4. N. M. V. Os, Ms. G. J. Daane, and T. A. B. M. Bolsman, *J. Colloid Interface Sci. 115*, 402 (1987).
5. P. K. Kilpatrick and W. G. Miller, *J. Phys. Chem. 88*, 1649 (1984).
6. T. Tahara, I. Satake, and R. Matuura, *Bull. Chem. Soc. Jpn. 42*, 1201 (1969).
7. Y. Chevalier and C. Chachaty, *Colloid Polym. Sci. 262*, 489 (1984).
8. R. Zana, *J. Colloid Interface Sci. 78*, 330 (1980).
9. E. J. R. Sudholter and J. B. F. N. Engberts, *J. Phys. Chem. 83*, 1854 (1979).
10. M. Krieg, M.-P. Pileni, A. M. Braun, and M. Grätzel, *J. Colloid Interface Sci. 83*, 209 (1981).
11. Z. Bedo, E. Berecz, and I. Lakatos, *Colloid Polym. Sci. 264*, 267 (1986).
12. S. Kucharski and J. Chlebicki, *J. Colloid Interface Sci. 46*, 518 (1974).
13. T. Kuwamura, in: *Structure/Performance Relationships in Surfactants* (M. J. Rosen, ed.), ACS Symp. Ser. No. 253, p. 27 (1984).
14. H. Matsumura, T. Watanabe, K. Furusawa, S. Inokuma, and T. Kuwamura, *Bull. Chem. Soc. Jpn. 60*, 2747 (1987).
15. J. M. Corkill, J. F. Goodman, P. Robson, and J. R. Tate, *Trans. Faraday Soc. 62*, 987 (1966).
16. J. H. Clint and T. Walker, *J. Chem. Soc. Faraday Trans. 1 76*, 946 (1975).
17. J. M. Corkill, J. F. Goodman, and T. Walker, *Trans. Faraday Soc. 63*, 759 (1967).
18. K. W. Herrmann, *J. Phys. Chem. 66*, 295 (1962).
19. W. L. Courchene, *J. Phys. Chem. 68*, 1870 (1964).
20. L. Benjamin, *J. Phys. Chem. 68*, 3575 (1964).
21. J. E. Desnoyers, G. Caron, R. DeLisi, D. Roberts, A. Roux, and G. Perron, *J. Phys. Chem. 87*, 1397 (1983).
22. Y. Moroi, E. Pramauro, M. Grätzel, E. Pelizzetti, and P. Tundo, *J. Colloid Interface Sci. 69*, 341 (1979).
23. K. W. Herrmann, J. G. Brushmiller, and W. L. Courchene, *J. Phys. Chem. 70*, 2909 (1966).
24. J. C. Lang and R. D. Morgan, *J. Chem. Phys. 73*, 5849 (1980).
25. M. Hanatani, K. Nishifuji, M. Futani, and T. Tsuchiya, *J. Biochem. 95*, 1349 (1984).
26. F. Yu and R. E. McCarty, *Arch. Biochem. Biophys. 61*, 238 (1985).
27. M. Okawauchi, M. Hagio, Y. Ikawa, G. Sugihara, Y. Murata, and M. Tanaka, *Bull. Chem. Soc. Jpn. 60*, 2718 (1987).
28. J. P. Kratohvil, W. P. Hsu, M. A. Jacobs, T. M. Aminabhavi, and Y. Mukunoki, *Colloid Polym. Sci. 261*, 781 (1983).
29. J. P. Kratohvil, *Hepatology 4*, 85S (1984).
30. P. Murkerjee, Y. Moroi, M. Murata, and A. Y. S. Yang, *Hepatology 4*, 61S (1984).
31. W. Clayton, *Theory of Emulsions*, 4th ed., p. 127, McGraw-Hill (Blakiston), New York (1943).
32. W. C. Griffin, *J. Soc. Cosmet. Chem. 1*, 311 (1949).
33. W. C. Griffin, *J. Soc. Cosmet. Chem. 5*, 249 (1954).
34. I. J. Lin, J. P. Friend, and Y. Zimmels, *J. Colloid Interface Sci. 45*, 378 (1973).
35. The Atlas HLB System, Atlas Chemical Ind. Inc. (now I. C. I. America Inc.), Wilmington, Del. (1963).
36. J. K. Haken, *Adv. Chromatogr. 17*, 163 (1979).
37. J. T. Davies, in: *Proceedings of 2nd International Congress of Surface Activity*, Vol. 1, p. 426, Butterworths, London (1957).

38. I. J. Lin, *J. Phys. Chem. 76*, 2019 (1972).
39. H. L. Greenwald, E. B. Kice, N. Kenly, and J. Kelly, *Anal. Chem. 33*, 465 (1961).
40. V. R. Huebner, *Anal. Chem. 34*, 488 (1962).
41. I. G. A. Fineman, *J. Am. Oil Chem. Soc. 46*, 296 (1969).
42. P. Becher and R. L. Birkmeier, *J. Am. Chem. Soc. 41*, 169 (1964).
43. J. Syzmanowski, A. Voelkel, J. Beger, and H. Merkwitz, *J. Chromatogr. 330*, 61 (1985).
44. J. Beger, H. Merkwitz, J. Syzmanowski, and A. Voelkel, *J. Chromatogr. 333*, 319 (1985).
45. P. A. Small, *J. Appl. Chem. 3*, 71 (1953).
46. H. Schott, *J. Pharm. Sci. 73*, 790 (1984).
47. W. C. Griffin, *Off. Dig. Fed. Paint Varn. Prod. Clubs 28*, 466 (1956).
48. R. C. Little, Ph.D thesis, Rensselaer Polytechnic Institute, Mic. 60-2689, Troy, N.Y. (1960).
49. Y. Barakat, M. El-Emary, L. Fortney, R. S. Schechter, S. Yiv, and W. H. Wade, *J. Colloid Interface Sci. 89*, 209 (1982).
50. K. Shinoda, M. Maekawa, and Y. Shibata, *J. Phys. Chem. 90*, 1228 (1986).
51. G. D. Miles and L. Shedlovsky, *J. Phys. Chem. 48*, 57 (1944).
52. K. Shinoda, T. Nakagawa, B. Tamamushi, and T. Isemura, *Colloidal Surfactants*, p. 42, Academic Press, New York (1963).
53. Y. Moroi, N. Nishikido, and R. Matuura, *J. Colloid Interface Sci. 50*, 344 (1975).
54. Y. Moroi, *J. Colloid Interface Sci. 122*, 308 (1988).
55. B. W. Barray and R. Wilson, *Colloid Polym. Sci. 256*, 251 (1978).
56. Y. Moroi, K. Motomura, and R. Matuura, *Bull. Chem. Soc. Jpn. 44*, 2078 (1971).
57. Y. Moroi, K. Motomura, and R. Matuura, *Bull. Chem. Soc. Jpn. 45*, 2697 (1972).
58. Y. Moroi, K. Motomura, and R. Matuura, *Bull. Chem. Soc. Jpn. 46*, 1562 (1973).
59. P. H. Elworthy and K. J. Mysels, *J. Colloid Sci. 21*, 331 (1966).
60. H. F. Huisman, *K. Ned. Akad. Wet. Proc. Ser. B 67*, 388 (1964).
61. P. J. Flory, *J. Am. Chem. Soc. 62*, 1561 (1940).
62. B. Weibull, in: *3rd International Congress of Surface Activity*, Cologne, Vol. 3, p. 121 (1960).
63. K. Nagase and Y. Sakagushi, *Kogyo Kagaku Zasshi 64*, 635 (1961).
64. H. Schott, *J. Pharm. Sci. 58*, 1521 (1969).
65. J. M. Corkhill, J. F. Goodman,and R. H. Ottewill, *Trans. Faraday Soc. 57*, 1627 (1961).
66. M. J. Schick, *J. Phys. Chem. 67*, 1796 (1963).
67. J. M. Corkhill, J. F. Goodman, and S. P. Harrold, *Trans. Faraday Soc. 60*, 202 (1964).
68. E. H. Crook, D. B. Fordyce, and G. F. Trebbi, *J. Phys. Chem. 67*, 1987 (1963).
69. J. Kelly and H. L. Greenwald, *J. Phys. Chem. 62*, 1096 (1958).
70. Nikko Chemicals Inc., A Summary of Research Papers using Homogeneous Polyethyleneglycol Alkyl Ether, Tokyo (1986).

3

Dissolution of Amphiphiles in Water

3.1. Introduction

Several terms are used to express relative ease of dissolution, such as *soluble at any proportion, very soluble, soluble, slightly or sparingly soluble,* and *insoluble.* Slightly or sparingly soluble materials have aqueous solubilities between about 10^{-2} and 10^{-6} mol $\cdot$ dm^{-3}; alcohols and carboxylic acids with n-alkyl chains from C_6 to C_{14} fall in this category. Surfactants can also be classed as barely soluble materials on the basis of their monomeric solubility or their concentration below the CMC, as is reasonably expected from the alkyl chain length of most surfactant molecules. However slight, aqueous solubility clearly indicates the presence of at least one hydrophilic group in a molecule. Particularly if this group is ionic, the degree of aqueous solubility depends strongly on the conditions of solution such as pH, ionic strength, the fraction of organic additives, and the temperature and pressure. Some sparingly soluble materials may be made quite soluble by altering the solution conditions, and the state of the dissolved molecules will vary accordingly.

Solubility is governed by the energy difference between the solid or liquid and dissolved states of materials, and especially by the stability of the solid state. The more stable a material, the smaller (and the more difficult to determine) is its solubility. However, the coexistence of a solute phase is thermodynamically significant: it reduces the degrees of freedom of the system by one. In any case, the differences between the physicochemical properties of a solution and those of the pure solvent come about by the dissolution of solutes into the solvent. This chapter discusses the process of dissolution from a thermodynamic standpoint.

3.2. Thermodynamic Parameters of Mixing

The extensive thermodynamic variables (Y) of systems consisting of a single phase are a homogeneous first-degree function in the independent variables $n_1, \ldots, n_c$, where $n_1, \ldots, n_c$ are the number of moles present in the system. This is known as *Euler's theorem*[1]:

$$Y(T, P, n_1, \ldots, n_c) = \sum_{i=1}^{c} n_i y_i (T, P, n_1, \ldots, n_c) \qquad (3.1)$$

where y_i is an intensive variable defined as

$$y_i = (\partial Y / \partial n_i)_{T,P,n_j} \qquad (3.2)$$

and constitutes the partial molar quantity of component i. The mean molar quantity of the mixture then becomes

$$y = Y \bigg/ \sum_{i=1}^{c} n_i = \sum_{i=1}^{c} x_i y_i \qquad (3.3)$$

For a two-component system, the partial molar quantities can be evaluated graphically from the mean molar quantity (Fig. 3.1). The slope at point

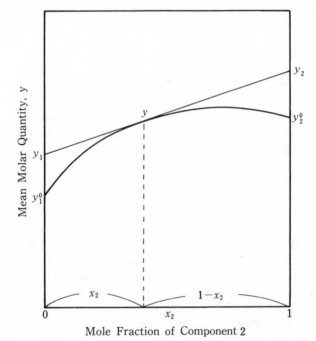

Mole Fraction of Component 2

Figure 3.1. Graphical method to determine the partial molar quantity from the mean molar quantity of a two-component system.

(x_2, y) is

$$(\partial y/\partial x_2)_{T,P} = y_2 - y_1 \tag{3.4}$$

and then, from Eqs. (3.3) and (3.4), the partial molar quantities y_1 and y_2 are given by

$$y_1 = y - x_2(\partial y/\partial x_2)_{T,P} \tag{3.5}$$
$$y_2 = y + (1 - x_2)(\partial y/\partial x_2)_{T,P} \tag{3.6}$$

Therefore, the two intercepts y_1 and y_2 of the tangential line turn out to be the partial molar quantities at the composition x_2. In addition, the following equation is set up among the partial molar quantities at constant temperature and pressure:

$$\sum_i n_i \, dy_i = 0 \tag{3.7}$$

When y_i is a chemical potential (μ_i), Eq. (3.7) reduces to the Gibbs–Duhem equation:

$$\sum_i n_i \, d\mu_i = 0 \quad \text{or} \quad \sum_i x_i \, d \ln a_i = 0 \tag{3.8}$$

at constant temperature and pressure, where a_i is the activity of component i.

The chemical potential—a partial molar Gibbs free energy of the component i in a solution—is expressed in the form

$$\mu_i = \mu_i^\circ(T, P) + RT \ln a_i = \mu_i^\circ(T, P) + RT \ln x_i\gamma_i \tag{3.9}$$

where $\mu_i^\circ(T, P)$ is the standard chemical potential in a symmetric reference system, which corresponds to the molar Gibbs free energy of a pure liquid component i at temperature T and pressure P. The thermodynamic concept μ_i° is very clear-cut for a solvent that is a major component of the solution (component 1). However, for solutes that are not necessarily liquid at the temperature and pressure specified and that, moreover, are present at low concentration, this definition of μ_i° is not appropriate. In the case where the mole fraction (x_1) of the solvent is very near unity, the solvent becomes ideal and satisfies the following expression:

$$\mu_1 = \mu_1^\circ + RT \ln x_1 \tag{3.10}$$

For a two-component system, the chemical potential of a solute (component 2) at low concentration can be derived from Eqs. (3.8) and (3.10):

$$d\mu_2 = (RT/x_2)\, dx_2 \qquad (3.11)$$

Integrating (3.11) with respect to x_2 yields

$$\mu_2 = RT \ln x_2 + \text{const} \qquad (3.12)$$

where const is the integration constant. The value of const has the following thermodynamic definition,[2] denoted by $\mu_2^{\ominus}$:

$$\text{const} = \lim_{x_1 \to 1} (\mu_2 - RT \ln x_2) = \mu_2^{\ominus}(T, P) \qquad (3.13)$$

The quantity $\mu_2^{\ominus}$ may be hypothetical, but the thermodynamic concept is very distinct, as is evidenced by Henry's law on solubility of gases in solution. Henry's law will be explained later by the lattice theory of solution.[3,4]

A thermodynamic system consisting of a pure component is divariant—i.e., has two degrees of freedom—and the partial molar quantity (y_i°) of the component i in a pure system is specified by temperature and pressure. Thus, the total extensive thermodynamic parameters before mixing are expressed by

$$Y^{\circ}(T, P, n_1, \ldots, n_c) = \sum_i n_i y_i^{\circ}(T, P) \qquad (3.14)$$

The change of the thermodynamic parameters due to mixing then becomes

$$Y^M(T, P, n_1, \ldots, n_c) = Y - Y^{\circ} = \sum_i n_i(y_i - y_i^{\circ}) \qquad (3.15)$$

where Y^M is the *function of mixing*.[1] Because the dissolution consists of the mixing of solvent and solute molecules, it can be handled by this function.

3.3. Lattice Theory of Solution

A liquid has a certain amount of short-range order around each molecule, when each molecule is taken as the origin for the radial distribution function of the molecules around it. This claim has been verified for liquid argon[5] and $CF_3Cl^{[6]}$ on the basis of x-ray diffraction studies. In other words, although a liquid does not have the kind of long-range order found in a crystal, each molecule in a liquid is surrounded by the nearest neighbor molecules, the average number of which is constant to within a small fraction. Therefore, the lattice theory of solution can be expected to be a very useful guide to the thermodynamic parameters of mixing of molecules in solutions. The problem here is that the number of the nearest neighbors and the lattice parameters are average quantities, rather than definite numbers as in a crystal. Therefore, the lattice model described below necessarily incorporates some assumptions.

Let us consider a two-component liquid mixture of components A and B, where N_A and N_B are the respective numbers of molecules, Z is the number of nearest neighbors, and E_{AB} is the pair-interaction energy between A and B molecules. Then, the number of pairs is

$$ZN_A = 2N_{AA} + N_{AB} \tag{3.16}$$

$$ZN_B = 2N_{BB} + N_{AB} \tag{3.17}$$

In order to obtain the total lattice energy $-\chi_A N_A$ due to the pair interactions, the following relations are assumed for the pure liquid A:

$$E_{AA}N_{AA} = -\chi_A N_A \tag{3.18}$$

$$E_{AA} = -2\chi_A/Z \tag{3.19}$$

where (3.19) comes from the relation $N_{AA} = ZN_A/2$. Likewise,

$$E_{BB} = -2\chi_B/Z \tag{3.20}$$

The interchange energy ω that is brought about by exchanging one molecule of A in pure liquid A for one molecule of B in pure liquid B arises from the breaking of two Z pairs A–A and B–B and formation of two Z pairs of A–B:

$$2ZE_{AB} - ZE_{AA} - ZE_{BB} = 2Z\omega \tag{3.21}$$

The canonical ensemble partition function then becomes

$$Q = q_A^{N_A}q_B^{N_B} \sum_{N_{AB}} g(N_A, N_B, N_{AB}) \exp(-E/kT) \tag{3.22}$$

where the total lattice energy is given by

$$E = N_{AA}E_{AA} + N_{BB}E_{BB} + N_{AB}E_{AB} \tag{3.23}$$

and q is a partition function from the internal degrees of freedom. Introducing Eqs. (3.16), (3.17), and (3.23) into Eq. (3.22), we obtain

$$Q = \{q_A \exp(\chi_A/kT)\}^{N_A}\{q_B \exp(\chi_B/kT)\}^{N_B}$$
$$\times \sum_{N_{AB}} g(N_A, N_B, N_{AB}) \exp(-\omega N_{AB}/kT) \qquad (3.24)$$

The Helmholtz free energy A then becomes

$$A = -N_A kT \ln q_A - N_A\chi_A - N_B kT \ln q_B - N_B\chi_B$$
$$-kT \ln\left\{\sum_{N_{AB}} g(N_A, N_B, N_{AB}) \exp(-\omega N_{AB}/kT)\right\} \qquad (3.25)$$

The molecules are assumed to be distributed among lattice sites in a completely random fashion in spite of nonzero molecular interaction (zeroth approximation). Hence, N_{AB} is given by

$$N_{AB} = ZN_A \times \{N_B/(N_A + N_B)\} \qquad (3.26)$$

which corresponds to ideal mixing and, moreover, to the ideal lattice statistics given by

$$\sum_{N_{AB}} g = (N_A + N_B)!/(N_A!N_B!) \qquad (3.27)$$

That is, the configurational free energy becomes independent of the lattice energy. Finally, Eq. (3.25) becomes

$$A = N_A\mu_A^{o,m} + N_B\mu_B^{o,m} + kT[N_A \ln\{N_A/(N_A + N_B)\}$$
$$+ N_B \ln\{N_B/(N_A + N_B)\}]$$
$$+ \omega ZN_AN_B/(N_A + N_B) \qquad (3.28)$$

where the molecular standard chemical potential $\mu_A^{o,m}$ and $\mu_B^{o,m}$ are

$$\mu_A^{o,m} = \frac{\partial}{\partial N_A}[-kT \ln\{q_A \exp(\chi_A/kT)\}^{N_A}]_{T,V} \qquad (3.29)$$

$$\mu_B^{o,m} = \frac{\partial}{\partial N_B}[-kT \ln\{q_B \exp(\chi_B/kT)\}^{N_B}]_{T,V} \qquad (3.30)$$

Then we have the molar chemical potentials in the form

$$\mu_A = N(\partial A/\partial N_A)_{T,V} = \mu_A^\circ + RT \ln x_A + NZ\omega x_B^2 \qquad (3.31)$$

$$\mu_B = N(\partial A/\partial N_B)_{T,V} = \mu_B^\circ + RT \ln x_B + NZ\omega x_A^2 \qquad (3.32)$$

where N is Avogadro's number. According to the definition of the standard chemical potential (3.13), $\mu_2^\ominus$ becomes

$$\mu_2^\ominus = \lim_{x_1 \to 1} (\mu_2^\circ + NZ\omega x_1^2) = \mu_2^\circ + NZ\omega \qquad (3.33)$$

This is the standard chemical potential in the asymmetric reference system. It clearly indicates that the standard chemical potential at infinite dilution is the sum of the partial free energy of a pure liquid at the specified temperature and pressure plus the interchange energy. The activity coefficients, γ_1 and γ_2, are, therefore, given from Eqs. (3.9),(3.31), and (3.32) in the forms:

$$RT \ln \gamma_1 = NZ\omega x_2^2 \qquad (3.34)$$

$$RT \ln \gamma_2 = NZ\omega x_1^2 \qquad (3.35)$$

In addition, the mean molar functions of mixing $y^M = Y^M/\sum_i n_i$ are derived from the above thermodynamic relations:

$$g^M = \sum_i x_i(RT \ln x_i \gamma_i) \qquad \text{(Gibbs free energy)} \qquad (3.36)$$

$$h^M = -\sum_i x_i\{RT^2(\partial \ln \gamma_i/\partial T)_P\} \qquad \text{(enthalpy)} \qquad (3.37)$$

$$s^M = -\sum_i x_i\{R \ln x_i \gamma_i + RT(\partial \ln \gamma_i/\partial T)_P\} \qquad \text{(entropy)} \qquad (3.38)$$

$$v^M = \sum_i x_i RT(\partial \ln \gamma_i/\partial P)_T \qquad \text{(volume)} \qquad (3.39)$$

The lattice theory of solution is derived from several idealized assumptions. The assumptions that components A and B are the same size and have the same number of nearest neighbors, for example, are not applicable to real solutions. The regular solution concept[7] of Hildebrand is more versatile: it takes into account a mixture of molecules of different sizes, where the principal idea is an ideal entropy of mixing at constant volume irrespective of heat. The activity coefficients in the form of (3.9) due to interaction between components A and B in a liquid mixture are derived by the following equations when the mixing term is expressed as a volume fraction:

$$RT \ln \gamma_A = v_A \phi_B^2(\delta_A - \delta_B)^2 \qquad (3.40)$$

$$RT \ln \gamma_B = v_B \phi_A^2 (\delta_A - \delta_B)^2 \qquad (3.41)$$

where ϕ_A and ϕ_B are the volume fractions of A and B, respectively:

$$\phi_A = n_A v_A / (n_A v_A + n_B v_B) \qquad (3.42)$$

$$\phi_B = n_B v_B / (n_A v_A + n_B v_B) \qquad (3.43)$$

and δ_A and δ_B are the solubility parameters, defined as

$$\delta_A = (E_A^v / v_A)^{1/2} \qquad (3.44)$$

$$\delta_B = (E_B^v / v_B)^{1/2} \qquad (3.45)$$

where E_A^v is the molar energy of evaporation of component A and v_A is its molar volume. The nonideality caused by interactions between A and B is included in the term $(\delta_A - \delta_B)^2$. Furthermore, Ben-Naim[8-10] derived the chemical potential of component A in a mixed solution of A and B by statistical mechanics, in the form

$$\mu_A = W(A|A + B; x_A) + kT \ln (\rho_A \Lambda_A^3 q_A^{-1}) \qquad (3.46)$$

where $W(A|A + B; x_A)$ is the coupling work of a molecule of A to the rest of the system composed of A and B with composition x_A; ρ_A is the number density of A; Λ_A is the momentum partition function of A; and q_A is the internal partition function. Equation (3.46) is rewritten as

$$\mu_A = \mu_A^{*,\rho} + kT \ln \rho_A \qquad (3.47)$$

where $\mu_A^{*,\rho}$ is the generalized standard chemical potential:

$$\mu_A^{*,\rho} = W(A|A + B; x_A) + kT \ln \Lambda_A^3 q_A^{-1} \qquad (3.48)$$

This is the standard chemical potential to be used in constructing the free energy of transfer of A from one phase to the other. If concentration is expressed as the mole fraction, then

$$\mu_A = \mu_A^{o,x} + kT \ln x_A \qquad (3.49)$$

for a dilute solution of A, where the standard chemical potential is

$$\mu_A^{o,x} = W(A|A + B; x_A) + kT \ln \rho_w \Lambda_A^3 q_A^{-1} \qquad (3.50)$$

According to the above three theories, the chemical potentials of a solute are composed of an intrinsic free energy of a pure component, an interaction energy between solute and solvent, and a concentration term. The quantities measured experimentally do not represent the chemcal potentials of the solute but rather their differences, and depend strongly on the interaction with solvent. The following section discusses how the solubility is determined.

3.4. Solubility

Let us extend the discussion on liquid–liquid mixtures to the solubility of a solid (component 2) in a solvent (component 1). The latter discussion is a special case of the former, but they differ in that a solid in equilibrium with a solution has very low energy compared with the liquid. Therefore, to apply the above discussion, we assume a supercooled liquid with a chemical potential far above that of a solid at the same temperature (Fig. 3.2). It is now possible to discuss the solubility of a solid using the mean

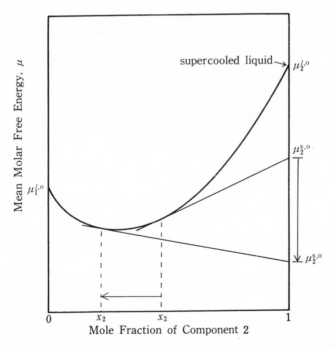

Figure 3.2. Change of mean molar free energy with composition of the solvent (1) and a supercooled liquid of the solute (2).

molar free energy curve of a two-component mixture. Because the value at an intercept μ_2^o of the tangential line of the curve with the ordinate at $x_2 = 1$ is the chemical potential or the partial molar free energy of a solute (see Fig. 3.1), the composition x_2 of the curve that represents equilibrium with the solid is the one whose tangential line reaches the free energy of a solid. In the composition range $0 \le x \le x_2$, the solid dissolves completely. For a more stable solid, the free energy of the supercooled liquid increases and the composition x_2 in equilibrium with the solid becomes less, resulting in a lower solutibility. Conversely, for a less stable solid the free energy of the supercooled liquid becomes lower, and if the interaction between solute and solvent is enhanced, the mean molar free energy curve becomes deeper. Both changes lead to a larger equilibrium composition x_2 and thus to a higher solubility of the solute. In general, therefore, a solute with a greater heat of fusion is less soluble, and a solute with a more exothermic heat of dissolution is more soluble.

When an excess solute phase(s) is a single component and is in equilibrium with a solution phase (component 1), the chemical potentials of the solute (component 2) are equal in both phases:

$$\mu_2^{s,o}(T, P) = \mu_2^l(T, P, x) \tag{3.51}$$

where the chemical potential μ_2^s of the solute phase is a function of T and P, and the chemical potential in a solution phase is a function of T, P, and the composition x_2 on the other hand. For an infinitesimal change of T along the solubility curve at constant pressure, we have the following equality from the differential change $d\mu_2^s = d\mu_2^l$:

$$-s_2^s \, dT = -s_2^l \, dT + (\partial \mu_2^l / \partial x_2)_{T,P} \, dx_2 \tag{3.52}$$

Equation (3.51) is rewritten as

$$h_2^s - Ts_2^s = h_2^l - Ts_2^l \tag{3.53}$$

From Eqs. (3.52) and (3.53), there results

$$\Delta h_2 / T = (\partial x_2 / \partial T)_P (\partial \mu_2^l / \partial x_2)_{T,P} \tag{3.54}$$

where $\Delta h_2 = h_2^l - h_2^s$. Introducing (3.9) into (3.54), we obtain

$$\Delta h_2 / RT^2 = (\partial \ln x_2 / \partial T)_P \{1 + x_2 (\partial \ln \gamma_2 / \partial x_2)_{T,P}\} \tag{3.55}$$

The heat of mixing for two components 1 and 2 is rewritten from (3.15) in the form

$$H^M = n_1(h_1 - h_1^\circ) + n_2(h_2 - h_2^\circ) \tag{3.56}$$

which is directly related to the integral heat of solution H^M/n_2 of component 2. The differential heat of solution Δh^{sol} with respect to solute component 2, on the otherhand, is defined as

$$\Delta h^{sol} = (\partial H^M/\partial n_2)_{T,P,n_1} = h_2 - h_2^\circ \tag{3.57}$$

because h_1 and h_2 are a zeroth order homogeneous function of n_1 and n_2 (Euler's theorem). Differentiation of (3.56) with respect to solvent component 1 leads to the differential heat of dilution Δh^{dil}. It turns out from (3.55) and (3.57) that the differential heat of solution can be evaluated from the change in solubility with temperature when the composition dependence of the activity coefficient is small.

3.5. Solubility of Weak Acids and Dissociation Constant

The aqueous solubility of most ionic surfactants depends on two intensive parameters, or on temperature at atmospheric pressure (see Chapter 6 for a detailed discussion). However, the total solubility of a sparingly soluble organic acid is more sensitive to pH than to temperature. This is because undissociated monomeric acids are in equilibrium with the dissociated forms, and the concentration ratio of dissociated to undissociated acid depends strongly on the solution pH. This pH dependence is represented by the *dissociation* or *acidity constant*.

When an excess solid or liquid phase of a monobasic acid (SH) coexists with an aqueous phase, the total solubility C_t, which is the sum of the concentration C_{S^-} of dissociated acid and the concentration C_{SH} of undissociated acid, depends upon the solution pH and can be expressed as

$$C_t = C_{SH} + C_{S^-} = C_{SH}(1 + K_a/\gamma_{S^-}a_{H^+}) \tag{3.58}$$

and the acidity constant K_a is given by

$$K_a = a_{S^-}a_{H^+}/C_{SH} \tag{3.59}$$

where the activity of the nonionic species is assumed to be equal to its concentration because its concentration is extremely low.[11,12] Because both C_{SH} and γ_{s^-} (an activity coefficient of S^-) remain almost constant at constant temperature and pressure, this equality predicts a linear relationship between C_t and $1/a_{H^+}$ (Fig. 3.3). Dividing the slope by the intercept of the line gives the value of k_a/γ_{s^-}. For a dibasic acid (SH_2), on the other hand, the total analytic concentration of the acid C_t is

$$C_t = C_{SH_2} + C_{SH} + C_S$$
$$= C_{SH_2}\{1 + (K_{a1}/\gamma_{SH})(1/a_{H^+}) + (K_{a1}K_{a2}/\gamma_S)(1/a_{H^+})^2\} \quad (3.60)$$

When the ionic strength of the solution is kept constant, or when the total concentration is low, every parameter except a_{H^+} of the right-hand side of (3.60) remains constant, and C_t changes with a second-power curve of $1/a_{H^+}$. If K_{a1} and K_{a2} differ greatly, (3.60) can be divided into two parts, one for pH $< pk_{a1}$, and the other for $pk_{a1} \ll$ pH $< pK_{a2}$. In the first region, a dibasic acid can be treated just like a monobasic acid (Fig. 3.4). The problem, however, is to evaluate K_{a1} and K_{a2} when K_{a1} is close to K_{a2}. From (3.60), the values of K_{a1} and K_{a2} can be obtained from the three coefficients of the graph $A + B(1/a_{H^+}) + C(1/a_{H^+})^2$ as

$$K_{a1} = \gamma_{SH}(B/A) \quad \text{and} \quad K_{a2} = (\gamma_S/\gamma_{SH})(C/B) \quad (3.61)$$

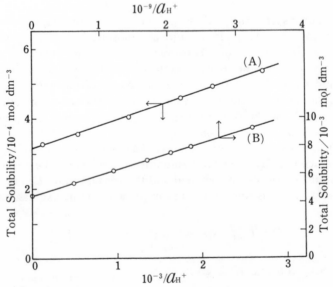

Figure 3.3. Plots of total solubility against $1/a_{H^+}$: (A) 1-naphthoic acid; (B) 2-naphthol.[12] (Reproduced with permission of Elsevier Science Publishers.)

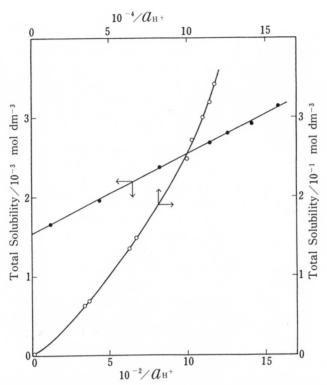

Figure 3.4. Plots of total solubility of diphenic acid against $1/a_{H^+}$[12]: (●) pH ≪ pK_{a1}; (○) pH < 5.5. (Reproduced with permission of Elsevier Science Publishers.)

Thus, both precise determination of the total concentration of the acid C_t in aqueous phases of different pH and successive measurement of three coefficients are necessary in order to estimate K_{a1} and K_{a2}. The activity coefficients γ_{SH} and γ_S can be estimated by the Debye–Hückel approximation. The key point of the solubility method is that the concentration of undissociated SH_n in the aqueous phase remains constant at constant temperature and pressure, regardless of the introduction of any other chemical species into the aqueous phase, so long as the total ionic strength is kept constant and low. This is the case because SH_n in the aqueous phase is in equilibrium with solid SH_n, i.e., the chemical potential of SH_n is kept constant at constant temperature and pressure.

We now consider the aqueous solubility and aggregation of sparingly soluble organic acids that are hydrophobic enough to form aggregates when the concentration of dissociated acid is increased by raising the pH (Fig. 3.5). The total equivalent concentration of monobasic acid departs from

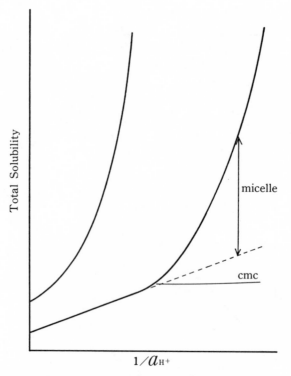

Figure 3.5. Total solubility change with pH for monobasic acid with long alkyl chains.

the straight line when aggregation begins. If we take the following stepwise associations for the dissociated acids (S) and counterions (G):

$$2S + m_1G \overset{K_1}{\rightleftharpoons} M_2$$

$$3S + m_2G \overset{K_2}{\rightleftharpoons} M_3$$

$$\cdots$$

$$iS + m_iG \overset{K_i}{\rightleftharpoons} M_i \tag{3.62}$$

then the total equivalent concentration of the dissociated acids C_e is expressed as

$$C_e = C_S + \sum_i iK_iC_s^iC_G^{m_i} \tag{3.63}$$

and the total molar concentration C_t is given by

$$C_t = C_s + \sum_i K_i C_s^i C_G^{m_i} \tag{3.64}$$

The concentration of undissociated monomeric acid is usually so small that it is neglected here. The mean aggregation number n of aggregates between the concentration C_s at which aggregation commences and the concentration C_e then becomes

$$n = (C_e - C_s)/(C_t - C_s) \tag{3.65}$$

The numerator of (3.65), which is an equivalent concentration used for aggregates, is easily determined from an extension line of the solubility curve. The problem is to obtain the total molar concentration—the denominator of (3.65)—from the solubility curve. Differentiation of (3.64) with respect to C_S gives

$$dC_t/dC_S = 1 + \sum_i iK_i C_S^{i-1} C_G^{m_i}$$

$$= C_e/C_S \tag{3.66}$$

where it is assumed that the counterion concentration is kept constant by the manipulated preparation of a buffered aqueous solution. Now the total molar concentration is calculated by the integration of (3.66) with respect to C_S:

$$C_t = \int_{C_S}^{C_t} (C_e/C_S)\, dC_S \tag{3.67}$$

The integrand is also evaluated at each monomer concentration from the solubility curve; after which the integration can be performed by a graphical method. This method is very useful for aggregates where the aggregation number is relatively small and increases with total equivalent concentration, for example, bile salts and dyes.[13,14] The case in which the counterions are not taken into account has already been discussed by Rossotti and Rossotti,[15] Mukerjee and Ghosh,[13,14,16,17] Petersen,[18] and French and Stokes.[19]

The solubility of conventional ionic surfactants has been under study for a long time and has been extensively described, as is discussed in more detail in Chapter 6. The reports that most strongly influenced succeeding work were those of Murray and Hartley[20] and McBain and Sierichs.[21] Recently, the solubility of surfactants in the presence of another counterion has come under study from the standpoints of the precipitation boundary[22-26] and phase change.[27]

References

1. I. Prigogine and R. Defay, *Chemical Thermodynamics*, Longmans, New York (1954).
2. J. G. Kirkwood and I. Oppenheim, *Chemical Thermodynamics*, McGraw-Hill, New York (1962).
3. E. A. Guggenheim, *Mixtures*, Oxford University Press, London (1952).
4. T. L. Hill, *An Introduction to Statistical Thermodynamics*, Addison-Wesley, Reading, Mass. (1962).
5. A. Eisenstein and N. S. Gingrich, *Phys. Rev. 62*, 261 (1942).
6. L. S. Bartell and L. O. Brockway, *J. Chem. Phys. 23*, 1860 (1955).
7. J. H. Hildebrand and L. Scott, *Regular Solutions*, Prentice-Hall, Englewood Cliffs, N. J. (1962).
8. A. Ben-Naim, *Water and Aqueous Solutions: Introduction to a Molecular Theory*, Plenum Press, New York 1(974).
9. A. Ben-Naim, *J. Phys. Chem. 82*, 792 (1978).
10. A. Ben-Naim and Y. Marcus, *J. Chem. Soc. 81*, 2016 (1984).
11. P. Mukerjee and Y. Moroi, *Anal. Chem. 50*, 1589 (1978).
12. Y. Moroi and R. Matuura, *Anal. Chim. Acta 152*, 239 (1983).
13. P. Mukerjee and A. K. Ghosh *J. Am. Chem. Soc. 92*, 6403 (1970).
14. A. K. Ghosh and P. Mukerjee, *J. Am. Chem. Soc. 92*, 6408 (1970).
15. F. J. C. Rossotti and H. Rossotti, *J. Phys. Chem. 65*, 926 (1961).
16. A. K. Ghosh and P. Mukerjee, *J. Am. Chem. Soc. 92*, 6413 (1970).
17. P. Mukerjee and A. K. Ghosh, *J. Am. Chem. Soc. 92*, 6419 (1970).
18. J. C. Petersen, *J. Phys. Chem. 75*, 1129 (1971).
19. H. T. French and R. H. Stokes, *J. Phys. Chem. 85*, 3347 (1981).
20. R. C. Murray and G. S. Hartley, *Trans. Faraday Soc., 31*, 183 (1935).
21. J. W. McBain and W. C. Sierichs, *J. Am. Oil Chem. Soc. 25*, 221 (1948).
22. R. Nemeth and E. Matijevic, *J. Colloid Interface Sci. 41*, 532 (1972).
23. N. Kallay, M. Pastuovic, and E. Matijevic, *J. Colloid Interface Sci. 106*, 452 (1985).
24. V. Hrust, N. Kallay, and D. Tezak, *Colloid Polym. Sci. 263*, 424 (1985).
25. N. Kallay, X.-J. Fan, and E. Matijevic, *Acta Chem. Scand. A40*, 257 (1986).
26. C. Noik, M. Baviere, and D. Defives, *J. Colloid Interface Sci. 115*, 36 (1987).
27. D. Tezak, F. Strajnar, O. Milat, and M. Stubicar, *Progr. Colloid Polym. Sci. 69*, 100 (1984).

4

Micelle Formation

4.1. Introduction

One of the most characteristic properties of amphiphilic molecules is their capacity to aggregate in solutions. The aggregation process depends, of course, on the amphiphilic species and the condition of the system in which they are dissolved. The abrupt change in many physicochemical properties seen in aqueous solutions of amphiphilic molecules or surfactants with long hydrophobic chains when a specific concentration is exceeded is attributed to the formation of oriented colloidal aggregates. The narrow concentration range over which these changes occur has been called the *critical micelle concentration* (CMC),[1-3] and the molecular aggregates that form above the CMC area are known as *micelles*. The difference between micellar colloids and other colloids is that micellar colloids are in dynamic equilibrium with monomers in the solution (see Section 4.7).

Micellar colloids represent dynamic association–dissociation equilibria. However, the theoretical treatment of micelles depends on whether the micelle is regarded as a chemical species or as a separate phase. The *mass action model*, which has been used ever since the discovery of micelles, takes the former point of view,[4-9] whereas the *phase separation model* regards micelles as a separate phase.[10-14] To apply the mass action model strictly, one must know every association constant over the whole stepwise association from monomer to micelle, a requirement almost impossible to meet experimentally. Therefore, this model has the disadvantage that either monodispersity of the micelle aggregation number must be employed or numerical values of each association constant have to be assumed.[15-20] The phase separation model, on the other hand, is based on the assumption that the activity[10,21-26] of a surfactant molecule and/or the surface tension[13,27-29] of a surfactant solution remain constant above the CMC. In reality, neither quantity remains constant,[8,30,31] so this model is also not strictly correct. Another new approach (discussed in more detail in Chapter 5) rests on the application of the thermodynamics of small systems[32] to micellar systems. In any case, over the past decade the nature of ionic micelles has been made clearer from studies of the activity of both surfactant ions[23-26,33,34] and counterions,[22,23,25,26,33,35] owing to development of new electrochemical techniques.

The variety of the theories on micelle formation results from the versatile properties of micelles. Thus, although a micelle may not have such a large aggregation number that it can be regarded as a phase in the usual sense, it still will have properties similar to those of a phase. At the same time, each micelle contains too many aggregated monomer molecules to be regarded as a chemical species, even a bulky chemical species.

In this respect, it is very instructive to consider the micellar solution system from the viewpoint of the phase rule. Figure 4.1 illustrates the changes in solubility and CMC of sodium tetradecyl sulfonate with temperature. If the micelle is regarded as a phase, three phases (intermicellar bulk phase, surfactant solid phase, and micellar phase) coexist along the solubility curve above T_k, and Gibbs's phase rule $f = C - P + 2$ (where f, C, and P are the number of degrees of freedom, component, and phase, respectively) gives only one degree of freedom, since the number of com-

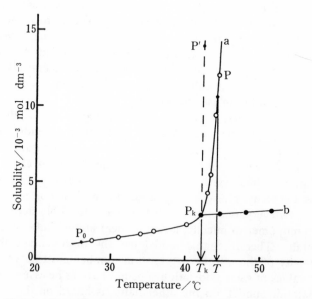

Figure 4.1. Changes of solubility (a) and CMC (b) with temperature for sodium tetradecyl sulfonate.[36] T_k = Krafft point. (Reproduced with permission of Academic Press.)

ponents is two (solvent water and surfactant). In other words, according to this model the solubility cannot change with temperature at constant pressure, because the solubility is automatically determined only by the pressure. On the other hand, if the mass action model is applied to a micelle formation, the solubility problem can be solved in a way consistent with the phase rule (see Chapter 6).[36] In addition, increase in the solubility observed above the point of micellization can be elucidated by the following semiquantitative discussion. Let us consider a simple association equilibrium between surfactant monomers (S) and micelles (M_n) of aggregation number n:

$$nS \underset{}{\overset{K_n}{\rightleftharpoons}} M_n \qquad (4.1)$$

The micellization constant K_n is therefore written as

$$K_n = [M_n]/[S]^n \qquad (4.2)$$

The equivalent concentration of surfactant (C_t) used for micelles then becomes:

$$C_t - [S] = nK_n[S]^n \qquad (4.3)$$

The ratio of the equivalent concentration at T to that at T_k is

$$\{C_t(T) - [S(T)]\}/\{C_t(T_k) - [S(T_k)]\} = ([S(T)]/[S(T_k)])^n \qquad (4.4)$$

where the K_n value is assumed constant because of the very small temperature range. The heat of dissolution obtained from the solubility change with temperature is about $100 \ kJ \cdot mol^{-1}$ for many ionic surfactants,[37-39] so the solubility change with temperature may be expressed roughly as:

$$[S(T_k + \Delta T)]/[S(T_k)] = 1 + 0.13 \times \Delta T \qquad (4.5)$$

For $\Delta T = 0.2°C$, the above ratios become 2.8 and 13.8 for $n = 50$ and $n = 100$, respectively.

It is evident that a small temperature increase brings about a large increase in solubility, and that the micelle aggregation number n has a very strong influence (Fig. 4.2). As is clear from the above discussion, the abrupt increase in the total solubility above T_k is due not to an increase in the solubility of the monomeric surfactant but rather to an increasing number of micelles. In addition, the treatment of micelles as a separate phase has turned out to be incorrect, whereas the mass action model is consistent not only with the phase rule but also with the solubility increase.

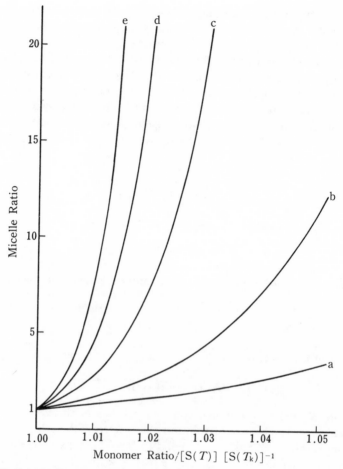

Figure 4.2. Solubility increase above the micelle temperature range depending on the aggrega-tion number of micelle (n). (a) $n = 25$; (b) $n = 50$; (c) $n = 100$; (d) $n = 150$; (e) $n = 200$. (Reproduced with permission of Academic Press.)

4.2. Shape and Structure of Micelles

Ever since McBain proposed the presence of molecular aggregates in soap solutions on the basis of the unusual changes in electrical conductivity observed with changing soap concentration,[40] the structure of micellar aggregates has been a matter of discussion. Hartley proposed that micelles are spherical with the charged groups situated at the micellar surface,[41] whereas McBain suggested that lamellar and spherical forms coexist.[42] X-ray studies by Harkins *et al.*[43] then suggested the sandwich or lamellar model.

Later, Debye and Anacker proposed that micelles are rod-shaped rather than spherical or disklike.[44] The cross section of such a rod would be circular, with the polar heads of the detergent lying on the periphery and the hydrocarbon tails filling the interior. The ends of the rod would almost certainly have to be rounded and polar. In 1956, Hartley's spherical micelle model was established by Reich[45] from the viewpoint of entropy, and the spherical form is now generally accepted as approximating the actual structure (Fig. 4.3).

The formation of micelles by ionic surfactants is ascribed to a balance between hydrocarbon chain attraction and ionic repulsion. The net charge of micelles is less than the degree of micellar aggregation, indicating that a large fraction of counterions remains associated with the micelle; these counterions form the *Stern layer* at the micellar surface. For nonionic surfactants, however, the hydrocarbon chain attraction is opposed by the requirements of hydrophilic groups for hydration and space. Therefore, the micellar structure is determined by an equilibrium between the repulsive forces among hydrophilic groups and the short-range attractive forces among hydrophobic groups. In other words, the chemical structure of a given surfactant determines the size and shape of its micelles.

The nature of micelles has been greatly clarified owing to recent progress in such research techniques as NMR, ESR, neutron scattering, and quasi-elastic light scattering. Neutron small-angle scattering experiments on sodium dodecyl sulfate and other ionic micelles support the basic Hartley model of a spherical micelle.[46-48] However, as the ion concentration is increased, the shape of ionic micelles changes in the sequence spherical–cylindrical–hexagonal–lamellar (Fig. 4.4).[49-52] For nonionic micelles, on the other hand, the shape seems to change from spherical directly to lamellar with increasing concentration.[53,54]

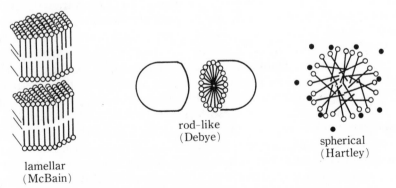

lamellar
(McBain)

rod-like
(Debye)

spherical
(Hartley)

Figure 4.3. Proposed shape and structure of the micelle.

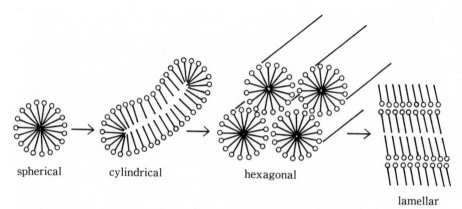

spherical   cylindrical   hexagonal

lamellar

Figure 4.4. Changes in micelle shape and structure with changing surfactant concentration.

Micelles of ionic surfactants are aggregates composed of a compressive core surrounded by a less compressive surface structure,[55] and with a rather fluid environment (of viscosity 8–17 cP for solubilized nitrobenzene in SDS and cetyltrimethylammonium bromide micelles).[56] Copper ions attached to micelles have essentially the same hydration shell near the micellar surface as in the bulk phase, and do not penetrate into the nonpolar part of the micelle.[57] In addition, it is known that the volume change caused by binding of divalent metal ions to micelles is very small.[58] The rate of rotation of the hydrated Na^+ ion at the micellar surface is unlikely to change by more than 35% upon adsorption from the bulk to the Stern layer of SDS micelles.[59]

In order to explain the relatively low degree of micelle ionization, Stigter and Mysels suggested that the micellar surface is rough,[60] and Stigter placed the hydrocarbon core–water interface at 0.4 to 1.2 Å from the center of the α-carbon atoms of ionic surfactants.[61] Furthermore, on the basis of NMR studies it has been proposed that the hydrocarbon tails do not penetrate into water across the micellar interface and that the first segments of the chains are nearly *trans* whereas the end segments have a conformation similar to that of a liquid hydrocarbon.[62] In other words, the penetration of water into the hydrocarbon micellar core must be very small, certainly less than one water molecule per surfactant molecule.[48] The micellar hydrocarbon core is virtually devoid of internal water, but NMR data suggest that substantial water/hydrocarbon contact occurs at the core interface.[63] Water/hydrocarbon contact is therefore limited to the micellar core surface. The water activity at the Stern layer of ionic micelles is not much less than in bulk water.[64] The rate of water reorientation at the ionic micellar surface is typically two to three times slower than in pure water, and the average lifetime of water molecules associated with micelles is between 6 and 37 ns.[65] For nonionic surfactants, the number of hydrating water molecules per oxyethylene unit was found to be three to five by gel filtration chromatography.[66] In apolar media, on the other hand, the micellar structures of nonionic surfactants are small cylindrical aggregates, and water molecules appear to be entrapped on the oxyethylene sites.[67]

4.3. Critical Micelle Concentration

As mentioned in Section 4.1, the narrow concentration range over which surfactant solutions show an abrupt change in physicochemical properties is called the critical concentration for the formation of micelles or critical micelle concentration. A variety of methods have been used to determine the CMC with or without additives in surfactant solutions.[2,27] Moreover, several definitions of CMC have been proposed. According to Corrin, the CMC is the total surfactant concentration at which a small and constant number of surfactant molecules are in aggregated form.[5] The number of aggregated surfactants at the CMC for a homologous series of surfactants is thus independent of hydrocarbon chain length. According to William et al,[68] the CMC is the concentration of surfactant solute at which the concentration of micelles would become zero if the micellar concentration continued to change at the same rate as it does at a slightly higher concentration of the solute. By this definition, the CMC is the concentration at which the two straight lines of solution properties below and above the CMC intersect each other. In 1955, taking the abrupt change of solution properties into consideration, Phillips defined the CMC as the concentration corresponding to the maximum change in a gradient in the solution property versus concentration (ϕ-C_t) curve[6]:

$$(d^3\phi/dC_t^3)_{C_t=\text{cmc}} = 0 \qquad (4.6)$$

where

$$\phi = \alpha[\text{S}] + \beta[\text{M}] \qquad (4.7)$$

α and β are proportionality constants, and [S] and [M] are the concentrations of the monomeric surfactant and micelle, respectively. Another definition, similar to that of Phillips, places the CMC at the point where

$$\{\partial(x_2 + x_m)/\partial x_d\}_{T,P} = 0.5 \quad \text{and} \quad x_d = x_2 + \bar{n}x_m \qquad (4.8)$$

where x_2 and x_m represent the mole fractions of nonmicellar and micellar surfactants, respectively, and $\bar{n}$ is the mean aggregation number of micelles.[69] This idea is easy to understand from the fact that the slope is 1 below the CMC and nearly zero above it. Finally, Israelachvili *et al.* defined the CMC as the concentration at which the analytical surfactant concentration in micelles equals the monomer concentration in bulk.[70] Of the above definitions, the one proposed by Phillips has been used most often.[71] Figure 4.5 illustrates the determination of the CMC from the electrical conductivity change with surfactant concentration.

Let us discuss the Phillips definition in more detail[72] because it is now in widespread use and seems to be the best for a number of reasons. Equation (4.6) is acceptable, but the property of a solution cannot be expressed as simply as by Eq. (4.7). Suppose that a surfactant molecule $G\nu_gS\nu_s$ is composed of ν_g counterions G with charge z_g, and ν_s surfactant ions S with

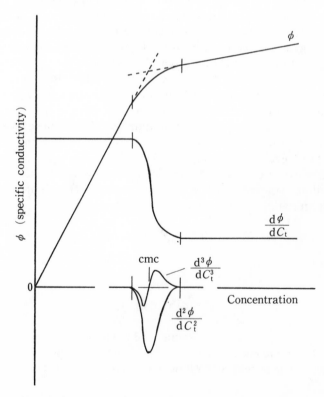

Figure 4.5. Schematic illustration of CMC determination according to the definition of Phillips.

charge z_s. To satisfy the condition of electrical neutrality,

$$z_g \nu_g + z_s \nu_s = 0 \tag{4.9}$$

As is well known, micelles are not monodisperse but polydisperse, and the following equilibria between counterions and surfactant ions can be given for micellization:

$$m_1 G + n_1 S \xrightleftharpoons{K_1} M_1$$

$$m_2 G + n_2 G \xrightleftharpoons{K_2} M_2$$

$$\cdots$$

$$m_i G + n_i S \xrightleftharpoons{K_i} M_i \tag{4.10}$$

Also, the electroneutrality of a solution holds:

$$z_g [G] + z_s [S] + \sum_i (z_g m_i + z_s n_i)[M_i] = 0 \tag{4.11}$$

where [G], [S], and [M_i] are the concentrations of counterions, surfactant ions, an micelles of aggregation number n_i, respectively. From (4.10), the micellization constant is written as

$$K_i = [M_i]/([G]^{m_i}[S]^{n_i}) \tag{4.12}$$

It is enlightening to examine the number of degree of freedom using Gibbs's phase rule ($f = C - P + 2 - r$) for the following discussion. The total number of components (C) is $i + 3$ (solvent, G, S, $M_1, \ldots, M_i$), the number of phases (P) is one (surfactant solution), and the number of equilibrium equations (r) is $i + 1$ [Eqs. (4.11) and (4.12)]. Therefore, the number of degrees of freedom is three, and the total surfactant concentration C_t determines the concentrations of every chemical species at constant temperature and pressure. The mass balances for counterions and surfactant ions, respectively, are expressed as

$$\nu_g C_t = [G] + \sum_i m_i [M_i] \tag{4.13}$$

$$\nu_s C_t = [S] + \sum_i n_i [M_i] \tag{4.14}$$

By analogy with (4.7), the solution property (ϕ) becomes a composite of the contributions from every chemical species:

$$\phi = \alpha_1 [S] + \alpha_2 [G] + \sum_i \beta_i [M_i] \tag{4.15}$$

where α_1, α_2, and β_i are the contribution factors of each chemical species. The magnitude of these properties depends on the solution properties used in determining the CMC. The solvent contribution is omitted here, since the surfactant concentration is relatively small and the solution property without a solvent contribution is usually employed.

With the help of (4.12), Eqs. (4.14) and (4.15) can be rewritten in terms of the concentrations of the monomeric surfactant ion and counterion:

$$\nu_s C_t = [S] + \sum_i n_i K_i [G]^{m_i}[S]^{n_i} \qquad (4.16)$$

$$\phi = \alpha_1[S] + \alpha_2[G] + \sum_i \beta_i K_i [G]^{m_i}[S]^{n_i} \qquad (4.17)$$

From (4.12)–(4.14) and (4.16), C_t and ϕ turn out to be functions of the common variables [S] or [G]. The derivative of ϕ with respect to C_t can be calculated through a common variable [S], and the CMC (C_t) can then be automatically determined from this [S] value.

It must be stressed that the CMC value obtained from (4.6) is a function of the contribution factors α_1, α_2, and β_i. In other words, the CMC depends on the solution properties employed in the determination and therefore differs with the method used. For this reason, measured CMC values define a narrow concentration range. The CMC values obtained from the solution properties mainly due to a monomeric surfactant contribution are found to be less than those due to a surfactant micelle contribution,[1-3] as can be seen in Fig. 4.6. In this case, random errors are taken into account for the CMC determination methods. For example, the CMC value obtained from surface tension measurements is less than that obtained from turbidity.[73] In the literature, however, CMCs have often been presented as definite concentrations,[74,75] especially since the appearance of a separation model for micellization.[13]

4.3.1. Monodisperse Micelles of Nonionic Surfactants

The simplest type of micelle is the monodisperse micelle of a nonionic surfactant for which $\nu_g = 0$, $z_s = 0$, and $n_i = n$. In this case, Eqs. (4.16) and (4.17) become, respectively,

$$C_t = [S] + nK_n[S]^n \qquad (4.18)$$

and

$$\phi = \alpha[S] + \beta K_n[S]^n \qquad (4.19)$$

cmc$/10^{-3}$ mol dm^{-3}

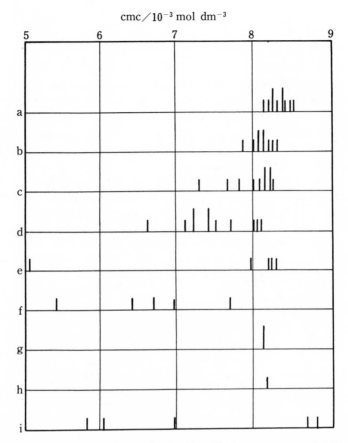

Figure 4.6. Variation of the sodium dodecyl sulfate CMC at temperatures between 20 and 30°C as obtained by various experimental methods.[3] a, specific conductivity; b, equivalent conductivity; c, other conductance; d, surface tension versus logarithm of concentration; e, other relationships between surface tension and concentration; f, absorbance; g, solubilization; h, light scattering; i, other methods such as refractive index, emf, vapor pressure, sound velocity, and viscosity.

The third derivative of ϕ with respect to C_t becomes

$$\mathrm{d}^3\phi/\mathrm{d}C_t^3 = (\beta - \alpha n)n(n-1)K_n[\mathrm{S}]^{n-3} \times \{(n-2)(1 + n^2 K_n[\mathrm{S}]^{n-1})$$

$$- 3n^2(n-1)K_n[\mathrm{S}]^{n-1}\}/(1 + n^2 K_n[\mathrm{S}]^{n-1})^5 \qquad (4.20)$$

The solution of $d^3\phi/dC_t^3 = 0$ gives

$$[\mathrm{S}] = \{(n-2)/(2n^3 - n^2)K_n\}^{1/(n-1)} \qquad (4.21)$$

and

$$C_t = \{(n - 2)/(2n^3 - n^2)K_n\}^{1/(n-1)} \times (2n^2 - 2)/(2n^2 - n) \quad (4.22)$$

The two important implications of these equations are that (1) the CMC (C_t) does not depend on the contribution factors, and (2) the micellization constant K_n can be estimated in terms of the values of the CMC and the aggregation number of the micelles. The former clear-cut arises from the monodispersity of micelles, and the CMC becomes independent of the method used for CMC determination.

Table 4.1 gives many K_n values evaluated using Eq. (4.22) for nonionic surfactants of the oxyethylene type whose CMC and aggregation number are known.[76-79] The K_n values increase with increasing aggregation number. From (4.21) and (4.22), the fraction of monomeric surfactants versus the total surfactant concentration at the CMC becomes

$$[S]/C_t = 1 - (n - 2)/(2n^2 - 2) \quad (4.23)$$

Figure 4.7 shows this fraction plotted against n. The monomeric fraction is

Table 4.1. Micellization Constant (K_n) of Nonionic Surfactants of Oxyethylene Type Based on Monodisperse Micelle[a]

Lipophile	Oxyethylene unit	CMC μmol·dm^{-3}	n	log K_n
1-Dodecanol	8	110	123	478
	12	93	81	318
	18	83	51	200
	23	91	40	154
Nonylphenol	15	110	80	309
	20	140	62	231
	30	185	44	157
	50	280	20	65
1-Tridecanol	10	125	88	335
	25	250	38	138
	22	196	28	97

[a] Reproduced with permission of the Chemical Society of Japan.[72]

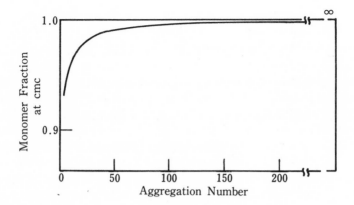

Figure 4.7. Monomer fraction at CMC with aggregation number of monodisperse micelle of nonionic surfactant.

found to be more than 99% at the CMC for common nonionic surfactant micelles with aggregation numbers larger than 50. Desnoyers *et al.* reached a different expression from a similar standpoint.[80]

4.3.2. Monodisperse Micelles of Ionic Surfactants

Let us now consider the monodisperse micelles of an ionic surfactant. From the electroneutrality of a solution as given by (4.11), the micelle concentration $[M_n]$ becomes

$$[M_n] = K_n[S]^n[G]^m = -x[S] - y[G] \qquad (4.24)$$

where

$$x = z_s/(z_g m + z_s n) \qquad (4.25)$$

and

$$y = z_g/(z_g m + z_s n) \qquad (4.26)$$

The equations corresponding to (4.16) and (4.17) then become, respectively,

$$\nu_s C_t = (1 - nx)[S] - ny[G] \qquad (4.27)$$

and

$$\phi = (\alpha_1 - \beta x)[S] + (\alpha_2 - \beta y)[G] \tag{4.28}$$

The third derivative of ϕ with C_t can likewise be obtained as

$$(1/\nu_s^3)(\mathrm{d}^3\phi/\mathrm{d}C_t^3) = (\alpha_2 - \beta y - \alpha_2 nx + \alpha_1 ny)$$

$$\times \{G'''(1 - nx - nyG')$$

$$+ 3nyG''^2\}/(1 - nx - nyG')^5 \tag{4.29}$$

where G', G'', and G''' are, respectively, the first, second, and third derivatives of $[G]$ with respect to $[S]$. The solution then becomes

$$G'''(1 - nx - nyG') + 3nyG''^2 = 0 \tag{4.30}$$

In this case, too, the monomer concentration at the CMC does not include any contribution factor at all.

It is evident that, in theory, the CMC does not depend on the method of CMC determination, regardless of whether ionic or nonionic micelles are involved, provided the micelles are monodisperse. From the experimental standpoint, however, there must be random errors for the CMC values. In other words, the CMC values coincide with one another, independent of the CMC determination method, whether the solution property on which it is based is due to the surfactant monomer or to the micelle. In reality, however, measured CMC values fall in a narrow concentration range.[2] This empirical finding indicates that the micelles formed in a surfactant solution are polydisperse, not monodisperse.

4.3.3. Polydisperse Micelles of Nonionic Surfactants

Let us take the simplest example in which micelles with two kinds of aggregation numbers (n_1 and n_2) are formed in a solution. In this case, (4.16) and (4.17) become, respectively,

$$C_t = [S] + n_1 K_1[S]^{n_1} + n_2 K_2[S]^{n_2} \tag{4.31}$$

and

$$\phi = \alpha[S] + \beta_1 K_1[S]^{n_1} + \beta_1 K_1[S]^{n_2} \tag{4.32}$$

Thus, from (4.31) and (4.32),

$$d^3\phi/dC_t^3 = N/(1 + n_1^2 K_1[S]^{n_1-1} + n_2^2 K_2[S]^{n_2-1})^5 \qquad (4.33)$$

$$
\begin{aligned}
N = {} & (\beta_1 - \alpha_1 n_1) n_1 (n_1 - 1) K_1 [S]^{n_1-3} \\
& \times \{n_1 - 2 + n_1^2 (1 - 2n_1) K_1 [S]^{n_1-1} + n_2^2 (1 + n_1 - 3n_2) K_2 [S]^{n_2-1}\} \\
& + (\beta_2 - \alpha_1 n_2) n_2 (n_2 - 1) K_2 [S]^{n_2-3} \\
& \times \{n_2 - 2 + n_1^2 (1 - 3n_1 + n_2) K_1 [S]^{n_1-1} + n_2^2 (1 - 2n_2) K_2 [S]^{n_2-1}\} \\
& + n_1 n_2 (n_1 - n_2)(\beta_1 n_2 - \beta_2 n_1) K_1 K_2 [S]^{n_1+n_2-4} \\
& \times \{n_1 + n_2 - 3 + n_1^2 (-2n_1 + n_2) K_1 [S]^{n_1-1} \\
& + n_2^2 (n_1 - 2n_2) K_2 [S]^{n_2-1}\}
\end{aligned}
\qquad (4.34)
$$

It turns out from (4.34) that the monomer concentration [S] satisfying $d^3\phi/dC_t^3 = 0$ depends on the contribution factors α_1, β_1, and β_2. That is, the CMC value obtained from the monomer concentration depends on the method of CMC determination. Thus, the difference in CMC values resulting from the method used is a systematic difference (by analogy with the difference between systematic errors and random errors). The micelle aggregation number in a real system is much more disperse than in the present case.[16-20,81,82] Therefore, in reality, the CMC obtained does depend on the method used for its determination, and the CMC value should be defined as a narrow concentration range even though each method yields a single CMC value.

As for the phase rule on a solubility curve, because an excess surfactant phase coexists in the system the number of phases is two. Thus, the number of degrees of freedom is two, as is clear from the above discussion, and therefore, the temperature determines the concentration of every chemical species at a specified pressure. This conclusion is in total agreement with the observation that the solubility is determined only by the temperature at atmospheric pressure. A CMC value should be a specific property for each surfactant. Methods that use a third component as an indicator are therefore not recommended, because addition of the third component may alter the structure and the stability of micelles in an unexpected way. In the case of surfactants for which the CMC cannot be determined without an indicator, it must be obtained by extrapolating the indicator concentration to zero.

4.4. Thermodynamics of Micelle Formation

The mass-action model should be verified before we discuss micelle thermodynamics. Recent progress in electrochemical techniques makes it possible to measure monomeric concentrations of surfactant ions and counterions, and determination of the micellization constant has become possible. The first equality of (4.24) has three parameters to be determined— K_n, n, and m, which are the most important factors for the mass-action model of micelle formation. For monodisperse micelles, the following equations result from (4.13) and (4.14), respectively:

$$[M_n] = (\nu_g C_t - [G])/m \qquad (4.35)$$

$$[M_n] = (\nu_s C_t - [S])/n \qquad (4.36)$$

or

$$m/n = (\nu_g C_t - [G])/(\nu_s C_t - [S]) \qquad (4.37)$$

From the logarithm of the first equality of (4.24), one obtains

$$\log [S] = -(m/n) \log[G] - (1/n) \log K_n + (1/n) \log [M_n] \quad (4.38)$$

The value of m/n is evaluated from a slope of the relation obtained by plotting log [S] against log [G] above the CMC, because $\log[M_n]$ is negligibly small compared to log K_n. The value of m/n is also obtained from Eq. (4.37) if [S] and [G] are available for each surfactant concentration. Equation (4.38) is rewritten with (4.35) and (4.36) as

$$\log[S] = -(m/n) \log\{[\nu_g - (m/n)\nu_s]C_t + (m/n)[S]\}$$

$$- (1/n) \log K_n + (1/n) \log[(\nu_s C_t - [S])/n] \qquad (4.39)$$

The three micellization parameters can be evaluated from three bulk concentrations of S at different surfactant concentrations.

If reference data on SDS[23] are used to evaluate m/n (Fig. 4.8), the three micellization parameters are log K_n = 230, n = 64, and m = 46.7 in units of molar concentration.[83] On the other hand, the bulk concentrations [S] and [G] can be evaluated using (4.39) at a given surfactant concentration (Fig. 4.9). The above numerical values yield excellent agreement between calculated and observed monomer concentrations at higher surfactant concentrations. However, lower parameter values are necessary to give good

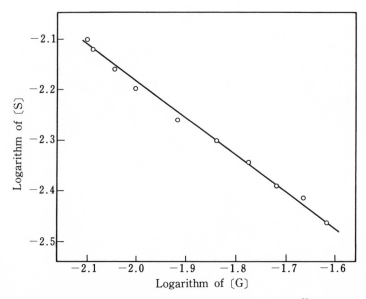

Figure 4.8. Relationship between logarithm of [S] and logarithm of [G].[83] (Reproduced with permission of Academic Press.)

agreement around the CMC: $\log K_n = 219$, $n = 62$, and $m = 45$. This means that the micellar equilibria displace to higher aggregation numbers with increasing total surfactant concentration, a result that is readily predicted from the polydispersity of micelles.

The important findings from the above calculations are: (1) the monomer concentrations of S and G at the CMC are more than 99% of the total surfactant concentration; (2) the CMC values due to $\Delta[S]/\Delta C_t$ and $\Delta[G]/\Delta C_t$ are identical; (3) the $\log[M_n]$ term is less than 3% of the $\log K_n$ term; and (4) the mass-action model agrees totally with micelle formation. On the basis of finding (2), the "premicelle formation" that has been suggested[9,84-88] cannot be allowed. In other words, the monomer concentrations [S] and [G] are essentially equal to the total surfactant concentration from just below the CMC downward. In addition, these facts imply that the slope of the logarithm of CMC plotted against the logarithm of total counterion concentration can be well approximated by the association degree of counterions of micelles (m/n):

$$\log \text{CMC} = -(m/n)\log[G] + \text{constant} \qquad (4.40)$$

A linear relationship between log CMC and log[G] has been observed in a number of experiments,[89-95] indicating that the approximation made above is correct.

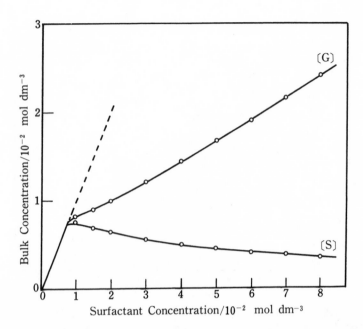

Figure 4.9. Change of bulk concentrations [S] and [G] with surfactant concentrations above the CMC.[83] O, reported value; —, theoretical value from $K_n = 230$, $n = 64$, and $m = 46.7$. (Reproduced with permission of Academic Press.)

Now that the mass-action model has been supported by a number of observations, we move to the thermodynamics of micelle formation based on this model. As would be predicted from the above discussion, micelle formation can be well expressed by a single association constant, even though the process strictly involves multiple association equilibria. The error is less than 5%, for example, for micelles having an aggregation number more than 50. For nonionic surfactants, the standard free energy change ΔG^0 per mole of surfactant molecules follows directly from the equilibrium constant and is given from (4.21) and [S] = C_t by

$$\Delta G^0 = -(RT/n) \ln K_n$$

$$= RT(1 - 1/n) \ln \mathrm{CMC} - (RT/n) \ln[(1 - 2/n)/(2n^2 - n)] \quad (4.41)$$

$$\Delta G^0 = RT \ln \mathrm{CMC} + (RT/n) \ln(2n^2) \qquad n > 50 \qquad (4.41')$$

The standard state is best chosen at an infinite dilution (Chapter 3) because the surfactant concentrations dealt with are usually dilute and because the observed change of thermodynamic variables is presumed to be equal to that on the basis of an infinite dilution state:

$$\Delta G^0 = (\mu_M^{\ominus} - n\mu_S^{\ominus})/n \tag{4.42}$$

From (4.41') we obtain the following equations for changes of thermodynamic variables:

$$\Delta H^0 = -RT^2[(\partial \ln \text{CMC}/\partial T)_p - 2(\partial n/\partial T)_p(1 - \ln 2n^2/2)/n^2] \tag{4.43}$$

$$\Delta S^0 = -(\partial RT \ln \text{CMC}/\partial T)_p - (R/n) \ln 2n^2$$
$$- 2RT(\partial n/\partial T)_p(1 - \ln 2n^2/2)/n^2 \tag{4.44}$$

$$\Delta V^0 = RT[(\partial \ln \text{CMC}/\partial P)_T - 2(\partial n/\partial P)_T(1 - \ln 2n^2/2)/n^2] \tag{4.45}$$

where ΔH^0, ΔS^0, and ΔV^0 are the corresponding enthalpy, entropy, and volume changes from (4.42).

For ionic surfactants, we assume a one-to-one electrolyte with monodisperse micelles. The relation between the equilibrium constant and CMC can be derived in principle from (4.30), but the differential equation is too complicated to solve in a clear-cut and closed way. We therefore apply the above findings to the micelle formation of ionic surfactants. In the vicinity of the CMC, Eqs. (4.27) and (4.28) can then be written respectively as

$$\nu_s C_t = [S] + nK_n[S]^{n+m} \tag{4.46}$$

and

$$\phi = (\alpha_1 + \alpha_2)[S] + \beta K_n[S]^{n+m} \tag{4.47}$$

The solution of $d^3\phi/dC_t^3$ leads to

$$1/K_n = [n(n + m)(2n + 2m - 1)/(n + m - 2)] \times [S]^{n+m-1} \tag{4.48}$$

The surfactant monomer concentration is almost equal to the CMC, and n is much larger than 1 for the usual process of micellization. The, (4.48) reduces to the approximate equation

$$1/K_n = 2n(n + m)(\text{CMC})^{n+m} \tag{4.49}$$

In this case, too, the standard free energy change per mole of surfactant ions follows from the equilibrium constant, and is given from (4.49) by

$$\Delta G^0 = (1 + m/n)RT \ln \text{CMC} + (RT/n) \ln[2n(n + m)] \tag{4.50}$$

where

$$\Delta G^0 = (\mu_M^\ominus - n\mu_S^\ominus - m\mu_G^\ominus)/n \tag{4.51}$$

The corresponding enthalpy, entropy, and volume changes are given respectively as

$$\Delta H^0/RT^2 = -(1 + m/n)(\partial \ln \mathrm{CMC}/\partial T)_p - [\partial(m/n)/\partial T]_p \ln \mathrm{CMC}$$

$$- \{\partial(1/n) \ln[2n(n + m)]/\partial T\}_p \tag{4.52}$$

$$\Delta S^0 = -(1 + m/n)(\partial RT \ln \mathrm{CMC}/\partial T)_p$$

$$- RT \ln \mathrm{CMC} \times [\partial(m/n)/\partial T]_p$$

$$- R\{\partial(T/n) \ln[2n(n + m)]/\partial T\}_p \tag{4.53}$$

$$\Delta V^0 = (1 + m/n)RT(\partial \ln \mathrm{CMC}/\partial P)_T$$

$$+ RT \ln \mathrm{CMC} \times [\partial(m/n)/\partial P]_T$$

$$+ RT\{\partial(1/n) \ln[2n(n + m)]/\partial P\}_T \tag{4.54}$$

A number of reports have been published concerning micelle formation from the statistical-thermodynamic[81,96-100] and thermodynamic[101-110] points of view. This is also the case for the changes of thermodynamic functions that occur when micelles form. If micelles were a separate phase, the second and the higher terms of the right-hand sides of (4.43) through (4.45) and (4.52) through (4.54) could be eliminated. However, these terms prove to be indispensable for obtaing a good fit between calculated and observed changes in thermodynamic variables.[111]

The volume change upon micellization (ΔV_m^0) can be estimated by two methods. One is based on the CMC change with pressure[112-115] given by (4.45) and (4.54), and the other depends on the change in the partial molar volume, as estimated from density measurements of the solutions below and above the CMC.[109,116-119] Figure 4.10 shows the effect of alkyl chain length on the partial molar volume. The volume increase that accompanies a transfer of the alkyl chain from the aqueous phase into the micelle is explained by a decrease in the number of hydrogen bonds of water molecules around the chain.[120] This idea is also the basis for the original concept of the entropy increase on micelle formation. In other words, the entropy increase on transfer of the alkyl chain from the aqueous environment to the liquid hydrocarbon phase is mainly due to the increase in entropy of the water around the alkyl chain caused by a transition from the state of "increased ice-likeness" to the normal state.

For the enthalpy change of micellization (ΔH^0), two procedures are also available: one based on the CMC change with temperature[121-124] by

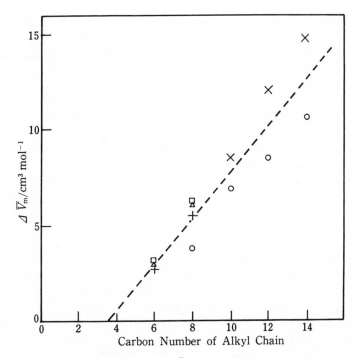

Figure 4.10. Partial molar volume change $\Delta \bar{V}_m$ on micelle formation plotted against alkyl chain length n_c.[117] O, $RN(CH_3)_3Br$; X, RSO_4Na; □, $RSO(CH_2)_2OH$; △, $RSO(CH_2)_3OH$; +, $RSO(CH_2)_4OH$; - - -, regression of $\Delta \bar{V}_m$ on n_c. (Reproduced with permission of the Royal Society of Chemistry.)

Eqs. (4.43) and (4.52), and the other based on either direct heat measurement, or heat measurement of dilution.[111,125-131] Values of ΔH^0 obtained by the two methods generally agree poorly for ionic surfactants[111,125,126] but much better for nonionic surfactants.[130] Furthermore, values have been found to decrease with an increase of excess counterion concentration[132-135] and also with surfactant alkyl chain length.[130,136-138] The latter change corresponds to the decrease in the free energy change of micelle formation with increasing alkyl chain length. The entropy change is not directly obtainable and is usually derived from the enthalpy change. Hence, the trend of ΔS^0 is similar to that of ΔH^0.

4.5. Counterion Binding to Micelles

Ionic surfactants are more difficult to aggregate in aqueous solutions than are nonionic surfactants of identical alkyl chain length, because higher concentrations are necessary to overcome the electrostatic repulsion between ionic head groups of ionic surfactants during aggregation. For nonionic surfactants, on the other hand, aggregation mainly due to hydrophobic attraction among alkyl chains is more feasible because the hydrophilic groups are easily separated from the water environment. As a result, CMC values of ionic surfactants are higher by one or more orders of magnitude than those of nonionic surfactants with the same hydrophobic groups, even though the micellization of both surfactant types is quite similar.

Many theoretical discussions on ionic micelle formation centered on taking electrostatic energy into consideration,[17,90,139-145] where the question was how to separate the hydrophobic energy from the electrical energy and how to evaluate the latter in connection with counterion binding to micelles. In Section 4.4 an ionic micelle was treated as a kinetic chemical species having bound counterions and with an electrical charge as a whole. In fact, as mentioned in Section 4.1, many reports have focused on the degree of counterion association with micelles. Table 4.2 summarizes the data on the binding of sodium ions to SDS micelles.[23] The degree of counterion binding ranges from 0.46 to 0.86, depending on the experimental technique employed, as would be expected from the lack of a definite distinction between bound and free counterions. This variability is just like the dependence of the CMC values on the method of determination.

Counterion binding values also differ depending on the model of micelle formation used: one model employs the assumed constancy of $[S] \times [G]$,[24,146] and the other that of $[S] \times [G]^{m/n}$.[23,25] Neither quantity can be assumed constant above the CMC, as can be seen from Fig. 4.11, which is calculated from the micellization parameters in Fig. 4.9.

Many theoretical discussions have centered on the degree of counterion binding to micelles,[140-147] because an understanding of the specific binding of counterions to micelles is a prerequisite for an understanding not only

Table 4.2. Degree of Counterion Binding to Micelles (m/n) of Sodium Dodecyl Sulfate[a]

m/n	Method of determination
0.86, 0.84, 0.78, 0.73, 0.63,	Electromotive force
0.85	Light scattering
0.82, 0.75, 0.73, 0.46	Mass-action model
0.82	Equilibrium dialysis
0.80, 0.74	Osmotic coefficient
0.72	Electrophoresis
0.50	Zeta-potential

[a] Reproduced with permission of the Chemical Society of Japan.[23,83]

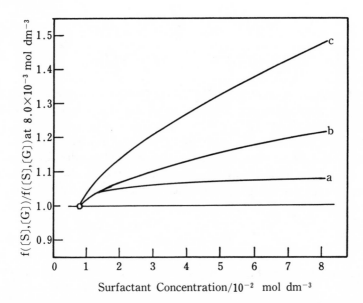

Figure 4.11. Change of functions $f([S], [G])$ with surfactant concentration.[83] a, $f = [S] \times [G]^{m/n}$; b, $f = ([S] \times [G])^{1/2}$; c, $f = [S] \times [G]$. (Reproduced with permission of Academic Press.)

of micellization but also of all kinds of aggregation processes in aqueous solutions. The theory developed by Evans and Ninham is one of the most reasonable.[145] Let us consider a spherical micelle immersed in a one-to-one electrolyte. The potential (ψ) around the micelle is represented from the electrolyte by the nonlinear Poisson–Boltzmann equation, which describes the distribution of ions at radius r around the micelle:

$$\mathrm{d}^2\psi/\mathrm{d}r^2 + (2/r) \times \mathrm{d}\psi/\mathrm{d}r = (8\pi n^0 e/\varepsilon) \times \sinh(e\psi/kT) \qquad (4.55)$$

where n^0 is the bulk electrolyte concentration, e the unit charge, ε the dielectric constant of the solvent, k Boltzmann's constant, and T the absolute temperature (see Section 7.1). The monomers are assumed to be completely dissociated and univalent. The boundary conditions are

$$\psi(r) \rightarrow 0 \qquad \text{as } r \rightarrow \infty \qquad (4.56)$$

$$\mathrm{d}\psi(r)/\mathrm{d}r = -4\pi\sigma/\varepsilon \qquad \text{at } r = R \qquad (4.57)$$

where σ is the surface charge density and R the distance from the center of the micelle to the surface. By introducing the following dimensionless variables into (4.55) and (4.57), we obtain (4.59) and (4.60):

$$y = e\psi/kT, \qquad x = \kappa r, \qquad \kappa^2 = 8\pi n^0 e^2/\varepsilon kT \qquad (4.58)$$

$$d^2y/dx^2 + (2/x) \times dy/dx = \sinh y \qquad (4.59)$$

$$dy/dx|_{x=\kappa R} = -4\pi\sigma e/\varepsilon\kappa kT \qquad (4.60)$$

The second term in (4.59), $(2/x)(dy/dx)$, describes a curvature correction due to a charged spherical micelle and drops out for a plane charged surface.

The next problem is to solve the differential equation (4.59) and to derive a useful closed form that agrees approximately with the numerical solution of (4.59).[148] For a first approximation, the micelle is assumed to be large enough for the micellar surface to be regarded as a plane. Then, (4.59) reduces to

$$d^2y/dx^2 = \sinh y \qquad (4.61)$$

From one of the above boundary conditions, the solution then becomes

$$dy/dx = -2\sinh(y/2) \qquad (4.62)$$

The sign of the right side depends on the charge of the micelle. Here, the micelle is assumed to be positively charged. By substituting (4.62) into the curvature term of (4.59) and integrating both sides with respect to y we have

$$\int_0^{y_0} (d/dy)[(dy/dx)^2/2]\,dy = \int_0^{y_0} \sinh y\,dy$$

$$+ 4\int_0^{y_0} (1/x)\sinh(y/2)\,dy \quad (4.63)$$

The main contribution to the second integral of the right side comes from the narrow integral region near $y = y_0$ or $x = kR$. For the second approximation, we remove this factor from the integral, giving

$$(dy/dx)^2/2|_{y=y_0} = 2\sinh^2(y_0/2) + (8/\kappa R)[\cosh(y_0/2) - 1] \quad (4.64)$$

Hence,

$$-(dy/dx)_{y=y_0} = 2\sinh(y_0/2)\{1 + (4/\kappa R)$$

$$\times [(\cosh(y_0/2) - 1)/\sinh^2(y_0/2)]\}^{1/2} \qquad (4.65)$$

The above analytic approximation is quite accurate for larger values of κR, and the error from the numerical solution of (4.59) is only 5% at $\kappa R = 0.5$. On the other hand, the surface charge density (σ) at a spherical micellar surface is given by

$$\sigma = (n - m)e/4\pi R^2 \qquad (4.66)$$

Introduction of (4.60) and (4.66) into (4.65) finally leads to

$$\frac{4\pi e^2(1 - \beta)}{\varepsilon \kappa a k T} = 2\sinh(y_0/2)\left\{1 + \frac{4}{\kappa R}\left[\frac{\cosh(y_0/2) - 1}{\sinh^2(y_0/2)}\right]\right\}^{1/2} \qquad (4.67)$$

where $\beta = m/n$ is the degree of the counterion binding to micelles and a is the area per surfactant molecule at the micellar surface. The surface potential y_0 is now a function of k, a and β. However, it depends mainly on κ because the other two variables remain almost constant, as is known from extensive data on the effect of counterion concentration. The counterion binding in polyelectrolyte solutions is similar: the apparent degree of dissociation of macroions remains almost constant irrespective of bulk counterion concentration after saturation of the binding.[149] These observations suggest that the saturation binding of counterions to polyelectrolytes applies not only to spherical micelles but also to rodlike or lamellar micelles and to polymeric macroions.

The gross features of the counterion binding or distribution between the kinetic micelle and the bulk solution can be understood in simplified electrostatic models like the one given above. However, the question is where to place the micellar surface. According to Linse *et al.*, the region lying between bulk and the micelle, as a kinetic entity, is relatively narrow, less than 0.5 Å deep.[144] Hence, the boundary can be said to be sharp. Numerous data on counterion binding lead to the following conclusions. The binding of alkali ions to anionic surfactants increases in the order $Li^+ < Na^+ < K^+ < Rb^+ < Cs^+$.[150] In other words, the CMC values decrease in the same order. This finding means that the bound counterions retain their primary hydration sheath, the diameter of which increases with decreasing ion size. This is also the case for cationic micelles. Binding of anionic counterions to cationic micelles increases in the order $F^- < Cl^- < Br^- < NO_3^- < I^-$.[151] The observed changes in CMC are then in line with the above discussion.[2,92,93,152] In addition, the counterion binding increases with increasing counterion hydrophobicity,[153-157] promoting micelle formation. For organic counterions, in particular, the binding of monovalent counterions with more than three methylene groups[154-156] and of divalent counterions with more than six methylene groups[39,158] increases steeply.

Lengthening the alkyl chain initially hinders micelle formation, but longer chains are markedly effective in lowering the CMC and in increasing the aggregation number, owing to enhanced hydrophobic interaction between the counterion and the micellar core. Counterion binding, in general, increases with increasing alkyl chain length for ionic surfactants,[14,159,160] and is decreased by addition of alcohols.[11,161,162] Divalent counterions, on the other hand, stimulate the growth of micelles, and this increase in growth is accompanied by both a decrease in the CMC of almost one order of magnitude compared with monovalent counterions for identical surfactant ions,[37,163-167] and an increase of counterion binding up to more than 0.9.[82,168] In addition, a model of micellar structure that includes counterions is now being elucidated by techniques of small-angle neutron scattering[46,51] and NMR.[169,170]

4.6. Size Distribution and Morphologic Alterations

As mentioned earlier, the aggregation number of micelles is not monodisperse but polydisperse. Therefore, their distribution is a matter of some concern. Let us take the model of micelle formation expressed by Eq. (4-1), where n is not definite but diffuse. Then, if μ_n and μ_1 are the chemical potentials of the micellar species composed of n monomers and the monomer, respectively, we have for the equilibrium between the monomers and any micellar species[102]

$$n\mu_1 = \mu_n \tag{4.68}$$

For solutions dilute enough to behave ideally, (4.68) yields

$$\ln x_n = -(\mu_n^{\ominus} - n\mu_1^{\ominus})/RT + n \ln x_1 \tag{4.69}$$

where $\mu_n^{\ominus}$ and $\mu_1^{\ominus}$ are the standard chemical potentials at infinite dilution of the micellar and monomer species, respectively, and x_1 is a solute mole fraction. By differentiating (4.69) with respect to $\ln x_1$ at constant temperature and pressure, we get

$$dx_n/d \ln x_1 = nx_n \tag{4.70}$$

Multiplying both sides of (4.70) by n^k, where k is a positive integer, and summing with respect to n leads to

$$d\left(\sum_n n^k x_n\right)\Big/d \ln x_1 = \sum_n n^{k+1} x_n \tag{4.71}$$

On the other hand, we can define the total solute fraction x_t and the total monomer fraction or the stoichiometric solute fraction x_m, expressed respectively by (4.72) and (4.73), as

$$x_t = x_1 + \sum_n x_n \tag{4.72}$$

$$x_m = x_1 + \sum_n nx_n \tag{4.73}$$

From (4.69), (4.72), and (4.73) we have

$$dx_t/dx_1 = x_m/x_1 \tag{4.74}$$

The mean micelle aggregation number n is written as

$$\bar{n} = \sum_n nx_n \bigg/ \sum_n x_n = (x_m - x_1)/(x_t - x_1) \tag{4.75}$$

where x_t can be evaluated from the integration of (4.74) with respect to x_1 (see Section 3.5). Thus, n is obtained when the change of x_1 with total surfactant concentration is experimentally available. Differentiating (4.75) with respect to $\ln x_1$ and rearranging with use of (4.71), we have

$$d\bar{n}/d \ln x_1 = \overline{n^2} - \bar{n}^2 \tag{4.76}$$

Then, the standard deviation σ of the aggregation number can also be obtained from (4.76) as

$$\sigma = (d\bar{n}/d \ln x_1)^{1/2} \tag{4.77}$$

From the above discussion, it becomes very important to determine the monomer fraction with total surfactant concentration (the stoichiometric solute fraction). Many size distributions of micelles have been proposed on various mathematical grounds.[19,20,139,171,172] Figure 4.12 illustrates the size distributions of counterion and surfactant ion obtained from counterion activity,[82] where the binding of counterions to a micelle is assumed to be equivalent to their solubilization to the micellar surface. The size distribution was found to be well expressed by the Poisson distribution, and the standard deviation of the distribution then becomes $\sqrt{\bar{n}}$.[173]

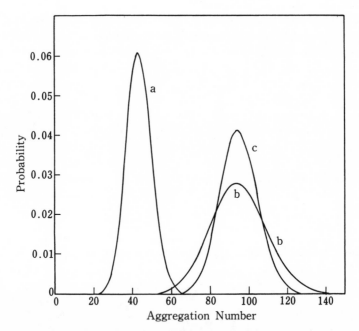

Figure 4.12. Size distribution of micelles with respect to counterion and surfactant ion of copper (II) dodecyl sulfate.[82] a, counterion (Cu^{2+}); b, surfactant ion; c, Poisson distribution ($n = 95$). (Reproduced with permission of the American Chemical Society.)

Some attempts have been made to estimate the aggregation number and the shape of micelles from the dependence of shape on micellar weight, on the area per polar group at the micellar surface, and on the molecular volume of the surfactant.[47,174–177] These studies found the micellar shape to resemble a sphere or an oblate or prolate ellipsoid not far removed from a sphere, and gave molecular weights or aggregation numbers approximately equal to those determined by other methods. These facts suggest that there is an upper limit to aggregation number and that the number of micelles increases with total surfactant concentration within a narrow size distribution.

On the other hand, the mean aggregation number should rise with total surfactant concentration, when considered from the standpoint of the mass-action model, as in (4.1). Indeed, some changes of colligative properties were found to take place at higher surfactant concentration far above the CMC; this transition point was named the *second CMC*[178] or *postmicellar transition.*[179] It has also become clear that the micellar aggregation number increases with increasing surfactant concentration[180] and with addition of salts.[44,181,182]

A morphologic alteration of micelles from a small sphere to a large prolate spherocylinder has been observed in a number of experiments after further addition of salts (Fig. 4.13).[183-191] These observations suggest that some type of energy barrier from an anticooperative region must be overcome for the shape transition from sphere to cylinder.[16] One way to overcome this barrier is to increase the chemical potential of monomers by increasing the total surfactant concentration, and the other is to decrease the barrier caused by reduced electrostatic repulsion by adding excess salts. The shape transition has also been observed after adding alcohols with a relatively short alkyl chain[192-195]; this effect may result from dilution of the micellar surface charge density owing to the solubilization of alcohols at the micellar palisade layer. Once a rodlike micelle is formed, the stepwise association of a monomer to a micelle merely leads to elongation of the micelle and becomes much easier to treat theoretically.

Many reports have considered the theoretical aspects of cylindrical micelles.[15,16,70,196-200] The theory proposed by Gelbert *et al.*[200] is presented here. Let us start with the statistical thermodynamics of a micellar solution.

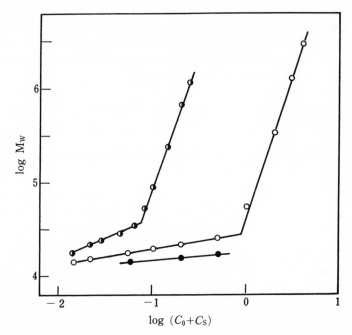

Figure 4.13. Relationship between logarithm of molecular weight (M_w) and logarithm of ionic strength [CMC (C_0) + salt concentration (C_s)].[185] ◑, dodecylammonium chloride at 30°C; ○, dodecyldimethylammonium chloride at 25°C; ●, dodecyltrimethylammonium chloride at 25°C (Reproduced with permission of Academic Press.)

Consider a surfactant solution containing N_t surfactant molecules in volume V at temperature T, where the micellar aggregates of different sizes coexist in equilibrium with free surfactant monomers in the solution. The mass conservation is now expressed by

$$\sum_n nN_n = N_t \tag{4.78}$$

where N_n denotes the number of micelles composed of n monomers including the free monomers $n = 1$. The partition function Q of the above canonical ensemble is given by

$$Q = \sum_{\{N_n\}} \prod_n (q_n^{N_n}/N_n!) \times (q_0^{N_0}/N_0!) \tag{4.79}$$

where the summation is over all distributions $\{N_n\}$ consistent with the conservation condition of (4.78); q_n is the partition function of a single n aggregate; q_0 is the partition function of a solvent molecule; and N_0 is the number of solvent molecules. The micelles are sufficiently dilute to be independent and indistinguishable, and q_n incorporates all external and internal degrees of freedom including solvent interaction. The partition function can be approximated by the maximum term (4.79),[201] and then the Helmholtz free energy A can be given by

$$A = -kTV \sum_n \rho_n[\ln(q_n/V) - \ln \rho_n + 1] - kT(N_0 \ln q_0 - \ln N_0!) \tag{4.80}$$

where $\rho_n = N_n/V$ is the number density of n micelles. The apparent mole fraction of n micelles is defined by

$$x_n = N_n \bigg/ \left(\sum_n nN_n + N_0\right) = \rho_n/\rho \tag{4.81}$$

where $\rho = (\sum_n nN_n + N_0)/V$ is the average number density of the solution. The stoichiometric solute mole fraction then becomes

$$x_t = \sum_n nx_n \tag{4.82}$$

The chemical potential of micelle n can be obtained from

$$\mu_n = (\partial A/\partial N_n)_{T,V,N_{j \neq n}} = \mu_n^0 + kT \ln x_n \tag{4.83}$$

where the standard chemical potential is given by

$$\mu_n^0 = -kT \ln (q_n/\rho V) \tag{4.84}$$

On the other hand, the most probable size distribution of micelles $\{N_n\}$ is determined by the condition that minimizes A, obeying the conservation constraint. From the standard minimization procedure we have

$$x_n^* = \exp[-(\mu_n^0 + \chi n)/kT] \tag{4.85}$$

where χ is the Lagrangian multiplier. The physical meaning of χ is apparent from the application of (4.85) to the free monomer: that is, $-\chi = \mu_1$. The size distribution results from the association–dissociation equilibria between micelles and monomers. The mass-action law is then applicable to the above equilibria as expressed by (4.68). We obtain the following equation from (4.68) and (4.83):

$$x_n = x_1^n \exp[n(\mu_1^0 - \mu_n^0)/KT] \tag{4.86}$$

This is another version of (4.69). Now we need to specify an explicit expression of μ_n^0 in order to calculate the micellar size distribution.

The shape change from sphere to spherocylinder—a cylinder capped by two end hemispheres—has been observed for many surfactant systems, and the rodlike micelles are the subject of the present discussion. The shape transition takes place by passing through the maximum spherical micelle whose aggregation number is m^* and whose standard chemical potential is $\mu_S^{0,*}(=m^*\bar{\mu}_S^{0,*})$. That is, the maximum micelle is the barrier that must be overcome for the shape transition from a sphere to a more stable long micelle. The standard chemical potential μ_n^0 of cylindrical micelle n is divided into two parts on the basis of their surface geometry, one for the two caps of m monomers and the other for the cylinder of $n - m$ monomers:

$$\mu_n^0 = \bar{n}_n^0 = m\bar{\mu}_s^0 + (n - m)\bar{\mu}_c^0$$

$$= n[\bar{\mu}_c^0 + (m/n)(\bar{\mu}_s^0 - \bar{\mu}_c^0)] = n(\bar{\mu}_c^0 + \alpha kT/n) \qquad n \geq m \tag{4.87}$$

where $\bar{\mu}_s^{0,*} > \bar{\mu}_s^0$ and $m^* > m$.[15,18,70,196] Substituting (4.87) into (4.85) or (4.86), we have

$$x_n = Bq^n \qquad n \geq m \tag{4.88}$$

where

$$B = \exp(-\alpha) = \exp[-m(\bar{\mu}_s^0 - \bar{\mu}_c^0)/kT] \qquad (4.89)$$

and

$$q = \exp[-(\bar{\mu}_c^0 + \chi)/kT] = x_1 \exp[(\mu_1^0 - \bar{\mu}_c^0)/kT] \qquad (4.90)$$

The value of B is less than unity, whereas that of q is nearly unity. The latter result is expected from the first equality of (4.90): namely, $\bar{\mu}_c^0$ approaches μ_1 as micelles advance to a separate phase.

The moment of x_n can now be evaluated by making use of Eq. (4.88). Defining the kth moment of x_n by

$$M_k = \sum_n n^k x_n = x_1 + B \sum_{n \geq m} n^k q^n \qquad (4.91)$$

we have

$$M_k = x_1 + B(q \, d/d \, q)^k [q^m/(1 - q)] \qquad (4.91')$$

For $k = 1$, M_1 becomes from (4.88)

$$M_1 = x_1 + B[mq^m/(1 - q) + q^{m+1}/(1 - q)^2] \qquad (4.92)$$

Here, the surfactant concentration is high enough to form rodlike micelles, and the x_1 is negligibly small. Under this condition, q can be assumed to be equal to unity, and we have the following approximate equations for the moments:

$$M_1 = x_t \simeq B/(1 - q)^2 \qquad (4.93)$$

or

$$q = 1 - (B/x_t)^{1/2} \qquad (4.93')$$

and

$$M_k = Bk!/(1 - q)^{k+1} \qquad (4.94)$$

From the above moments, the number-averaged aggregation number of micelles n and the corresponding standard deviation σ are respectively given by

$$\bar{n} = (M_1 - x_1)/(M_0 - x_1) = (x_t/B)^{1/2} \tag{4.95}$$

and

$$\sigma = (\overline{n^2} - \bar{n}^2)^{1/2} = (x_t/B)^{1/2} \tag{4.96}$$

On the other hand, we hve the following equations for the weight-averaged aggregation number $\langle n \rangle$ and the standard deviation σ_w:

$$\langle n \rangle = (M_2 - x_1)/(M_1 - x_1) = 2(x_t/B)^{1/2} \tag{4.97}$$

$$\sigma_w = (\langle n^2 \rangle - \langle n \rangle^2)^{1/2} = \langle n \rangle/\sqrt{2} \tag{4.98}$$

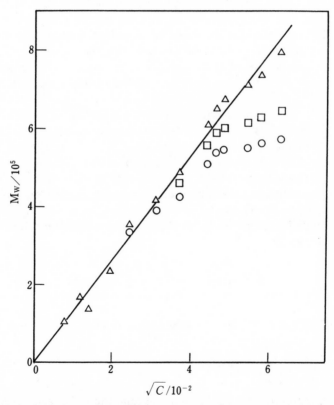

Figure 4.14. Variation in micellar molecular weights (M_w) of hexadecyltrimethylammonium bromide with $\sqrt{C}$.[15] $\bigcirc$, uncorrected molecular weights; $\square$, corrected for dissymmetry; $\triangle$, corrected for dissymmetry and second virial coefficients. (Reproduced with permission of the American Chemical Society.)

The weight-averaged aggregation number increases in proportion to the square root of total surfactant concentration, as has been observed experimentally (Fig. 4.14).

4.7. Kinetics of Micelle Formation

Micellar colloids are in a dynamic association–dissociation equilibrium, and the kinetics of micelle formation have been investigated for a long time.[202-209] In 1974, a reasonable explanation of the experimental results was proposed by Aniansson and Wall,[207] and this conception has been accepted and used ever since. The rate of micelle dissociation can be studied by several techniques, such as stopped flow,[210,211] pressure jump,[205,212] temperature jump,[203,213] ultrasonic absorption,[206,214,215] NMR,[216] and ESR.[217] The first three methods depend on tracing the process from a nonequilibrium state brought about by a sudden perturbation to a new equilibrium state—the relaxation process. The last two methods, on the other hand, make use of the spectral change caused by changes in the exchange rate of surfactant molecules between micelle and intermicellar bulk phase.

The relaxation experiments revealed two processes for micellar dissociation. The first of these is very fast, with a relaxation time of less than 10^{-5} s at surfactant concentrations around the CMC,[203,217,218] and results from an association–dissociation reaction of monomers to and from micelles. The other is a slower process, with a relaxation time of more than 10^{-3} s,[205] and is associated with a micelle formation–dissolution equilibrium.

The relaxation process around equilibrium obeys first-order reaction kinetics regardless of the order of reaction. Let us consider the following reaction, for example:

$$n_A A + n_B B + \ldots \underset{k_b}{\overset{k_f}{\rightleftharpoons}} n_C C + n_D D + \ldots \qquad (4.99)$$

At equilibrium the forward reaction rate v_e is equal to the back reaction rate v_b

$$v_e = k_f [A_e]^{n_A} \times [Be]^{n_B} \times \cdots = k_b [Ce]^{n_C} \times [De]^{n_D} \times \cdots \qquad (4.100)$$

where $[i_e]$ represents the equilibrium concentration of species i. When the equilibrium system is transformed to a nonequilibrium state $[\Delta i]$ by some perturbation, the system relaxes and adopts a new equilibrium state. The time required for the system to relax is the relaxation time τ. The reaction rate v_r to the new equilibrium state is given by

$$v_r = k_f([A_e] + [\Delta A])^{n_A} \times ([B_e] + [\Delta B])^{n_B} \times \cdots$$

$$- k_b([C_e] + [\Delta C])^{n_C} \times ([D_e] + [\Delta D])^{n_D} \times \cdots \qquad (4.101)$$

Introducing the extent of reaction

$$\Delta \xi = -[\Delta A]/n_A = -[\Delta B]/n_B = \cdots = [\Delta C]/n_C = [\Delta D]/n_D = \cdots \qquad (4.102)$$

into (4.100) and neglecting the second- and higher-order terms of $\Delta \xi$, we have

$$v_r = d(\Delta \xi)/dt = -v_e\{n_A^2/[A_e] + n_B^2/[B_e]$$

$$+ \cdots + n_C^2/[C_e] + n_D^2/[D_e] + \cdots\} \Delta \xi \qquad (4.103)$$

or

$$v_r = -(1/\tau) \Delta \xi \qquad (4.103')$$

where

$$1/\tau = \left[\sum_i (n_i^2/[i_e]) \right] v_e \qquad (4.104)$$

Equation (4.104) indicates that the relaxation reaction always obeys first-order decay kinetics.

Aniansson and Wall proposed the following kinetics of micelle formation, which are now generally accepted and have been adopted as the basis for the relaxation process.[207,219] Micellization is expressed by the stepwise association–dissociation equilibria

$$S_1 + S_1 \underset{\overleftarrow{k_2}}{\overset{\vec{k}_2}{\rightleftharpoons}} S_2$$

$$S_2 + S_1 \underset{\overleftarrow{k_3}}{\overset{\vec{k}_3}{\rightleftharpoons}} S_3$$

$$\cdots$$

$$S_{n-1} + S_1 \underset{\overleftarrow{k_n}}{\overset{\vec{k}_n}{\rightleftharpoons}} S_n \qquad (4.105)$$

where S_1 refers to the surfactant monomer, S_2 to the dimer, and so on. For ionic surfactants, S_1 denotes the surfactant ion. Counterions need not be taken into account because they are so mobile that they can be assumed to adjust almost immediately to the movement of surfactant ions. By introducing the relative deviation from equilibrium (ξ_n)

$$\xi_n = ([S_n] - [S_n^e])/[S_n^e] \tag{4.106}$$

there results the following kinetic equation:

$$d[S_n]/dt = [S_n^e] \times d\xi_n/dt = \vec{k}_n[S_1][S_{n-1}] + \overleftarrow{k}_{n+1}[S_{n+1}]$$
$$- \overleftarrow{k}_n[S_n] - \vec{k}_{n+1}[S_1][S_n] \tag{4.107}$$

where $[S_n]$ and $[S_n^e]$ denote the concentration of the n-mer micellar species and its equilibrium value, respectively. Using the equilibrium condition

$$\vec{k}_n[S_1^e][S_{n-1}^e] = \overleftarrow{k}_n[S_n^e] \tag{4.108}$$

we obtain

$$[S_n^e] \times d\xi_n/dt = \overleftarrow{k}_{n+1}[S_{n+1}^e][\xi_{n+1} - \xi_n(1 + \xi_1) - \xi_1]$$
$$- \overleftarrow{k}_n[S_n^e][\xi_n - \xi_{n-1}(1 + \xi_1) - \xi_1] \qquad n = 2, 3, \ldots \tag{4.109}$$

and for $n = 1$

$$[S_1^e] \times d\xi_1/dt = 2\overleftarrow{k}_2[S_2^e][\xi_2 - 2\xi_1 - \xi_1^2)$$
$$+ \sum_{n=3}^{\infty} \overleftarrow{k}_n[S_n^e][\xi_n - \xi_{n-1}(1 + \xi_1) - \xi_1] \tag{4.110}$$

The second term of the right side of (4.109) is the net increase in unit time of the number of aggregates of size n from the previous step:

$$J_n = \vec{k}_n[S_1][S_{n-1}] - \overleftarrow{k}_n[S_n]$$
$$= -\overleftarrow{k}_n[S_n^e][\xi_n - \xi_{n-1}(1 + \xi_1) - \xi_1] \tag{4.111}$$

At small deviations from equilibrium, the term $\xi_1\xi_{n-1}$ can be neglected:

$$J_n = -\overleftarrow{k}_n[S_n^e](\xi_n - \xi_{n-1} - \xi_1) \tag{4.111'}$$

Then, (4.107) or (4.109) is rewritten as

$$d[S_n]/dt = J_n - J_{n+1} \tag{4.112}$$

These micellar kinetics are strikingly analogous, in terms of the reaction flow in the aggregation space $\{n\}$, to one-dimensional diffusion through a tube of varying cross section: n corresponds to the space coordinate, ξ_n to the concentration, $[S_n^e]$ to the cross-sectional area, and $\bar{k}_n$ to the diffusion constant. Since $[S_n^e]$ is large at the monomer end and in the micellar region, the mass transfer tube has two side ends connected by a very narrow waist. This narrow waist corresponds to the intermediate-size aggregates of very low concentration. The mass transport from one end to the other of the tube can be expected to be much slower than the rapid attainment of a pseudoequilibrium within the two ends, the micellar and monomer regions. This analogy gives a good simplified picture of the presence of two relaxation times, one corresponding to the rapid monomer exchange between the micellar phase and the bulk solvent (τ_1) and the other to the slow change in the total number of micelles or the micelle formation–dissolution reaction (τ_2). Hence, the nonequilibrium state created by the perturbation is followed by an initial short period of adjustment in both ends, after which the main process of a pseudostationary flow from one end to the other end takes place.

The following schematic illustration will be used for the subsequent discussion of the mass transport. The $\{n\}$ space is split into three parts, $1 \leq n \leq n_1$, $n_1 + 1 \leq n \leq n_2$, and $n_2 + 1 < n$ (Fig. 4.15). The diffusions

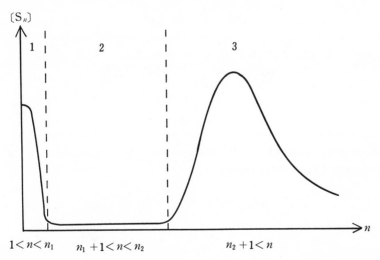

Figure 4.15. Division of aggregation space into monomer and oligomers (1), intermediate-size aggregates (2), and micellar aggregates (3).

$\tilde{k}_n[S_n^e]$ are assumed to be much larger in the first and third parts than some of the diffusion in the second part, and the amount of surctants in the second part,

$$\sum_{n>n_1}^{n_2} n[S_n] = \sum_{n>n_1}^{n_2} n(1 + \xi_n)[S_n^e]$$

is assumed to be negligible compared to the surfactant in the other two parts. It is also assumed that the deviation from equilibrium is so small that (4.111') is applicable for all n. Then, in the first and third parts

$$\xi_n - \xi_{n-1} - \xi_1 = -J_n/\tilde{k}_n[S_n^e]) \simeq 0 \qquad (4.113)$$

whereas in the second part

$$\xi_n - \xi_{n-1} - \xi_1 = -J/(\tilde{k}_n[S_n^e]) \qquad (4.113')$$

where J is the practically n-independent value of J_n during the pseudo-stationary phase considered. Because of the very low concentration in the second part, even a small difference between J_n and J_{n+1} would lead to large and rapid temperature variations, which is not the case for the pseudo-stationary phase. Summing (4.113) from $n = 2$ to n_1, (4.113') from $n = n_1 + 1$ to n_2, and (4.113) from $n = n_2 + 1$ to n'', we obtain

$$\xi_{n''} = n''\xi_1 - RJ \qquad (4.114)$$

where

$$R = \sum_{n=n_1+1}^{n_2} 1/(\tilde{k}_n[S_n^e]) \qquad (4.115)$$

R being a resistance formed by series connection of individual resistances $1/(\tilde{k}_n[S_n^e])$. For $n' \leq n_1$, we have similarly $\xi_{n'} = n'\xi_1$.

From the material balance we have

$$\sum_{n=1}^{\infty} n\xi_n[S_n^e] = 0 \qquad (4.116)$$

From (4.116), under the assumption of a negligible amount of surfactant in the second part, there results

$$\sum_{n'=1}^{n_1} n'[S_{n'}^e]\xi_{n'} + \sum_{n''=n_2+1}^{\infty} n''[S_{n''}^e]\xi_{n''} = 0 \qquad (4.117)$$

By introducing into $\xi_{n'}$, and $\xi_{n''}$ of (4.117) the corresponding relations, we obtain

$$J = (\overline{n_1^2}C_1 + \overline{n_3^2}C_3)\xi_1/ RC_3\bar{n}_3 = -(\overline{n_1^2}C_1 + \overline{n_3^2}C_3)m_3/(RC_1C_3\overline{n_1^2}\bar{n}_3) \qquad (4.118)$$

where

$$m_3 = \sum_{n > n_2} n\xi_n[S_n^e] \tag{4.119}$$

$$C_1 = \sum_{n=1}^{n_1} [S_n^e], \qquad C_3 = \sum_{n > n_2} [S_n^e] \tag{4.119'}$$

$$\overline{n_1^2} = \sum_{n=1}^{n_1} n^2[S_n^e]/C_1, \qquad \bar{n}_3 = \sum_{n > n_2} n[S_n^e]/C_3,$$

$$\overline{n_3^2} = \sum_{n > n_2} n^2[S_n^e]/C_3 \tag{4.119''}$$

To derive a differential equation for the process, we note that, on average, each micelle added to the third part increases the excess amount of material in that part by the amount

$$\partial m_3 / \partial\left(\sum_{n > n_2} \xi_n[S_n^e] \right)$$

so that

$$dm_3/dt = \left\{ \partial m_3 / \partial\left(\sum_{n > n_2} \xi_n[S_n^e] \right) \right\} \times J \tag{4.120}$$

The denominator of the right side of (4.120) is rewritten from (4.114) as

$$\sum_{n > n_2} \xi_n[S_n^e] = \xi_1 \sum_{n''} n''[S_{n''}^e] - RJ \sum_{n''} [S_{n''}^e] \tag{4.121}$$

Substituting (4.118) and (4.119) into (4.121) we have

$$\sum_{n > n_2} \xi_n[S_n^e] = (\overline{n_1^2}C_1 + \sigma^2 C_3)m_3 / \overline{n_1^2}\bar{n}_3 C_1 \tag{4.122}$$

where $\sigma^2 = \overline{n_3^2} - \bar{n}_3^2$, the variance of the micellar size distribution. Introducing (4.118) and (4.122) into (4.120) we have

$$dm_3/dt = -(1/\tau_2)m_3 \tag{4.123}$$

and

$$1/\tau_2 = (1/RC_3) \times [(\overline{n_1^2}C_1 + \overline{n_3^2}C_3)/(\overline{n_1^2}C_1 + \sigma^2 C_3)] \tag{4.124}$$

where τ_2 is the relaxation time for the slow process due to the micelle formation-dissolution reaction. Equation (4.124) can be further simplified by the following assumptions:

$$\overline{n_1^2} = 1 \quad \text{or} \quad \overline{n_1^2}C_1 = [S_1^e] \tag{4.125}$$

$$\bar{n}_3 C_3 = S_{\text{tot}} - [S_1^e] = S_{\text{mic}} \tag{4.126}$$

Thus, for a relatively small size distribution we have

$$\overline{n_3^2}C_3 = (\bar{n}_3^2 + \sigma^2)C_3 \simeq \bar{n}_3^2 C_3 = \bar{n}S_{\text{mic}} \tag{4.127}$$

where n is the mean micellar aggregation number. When S_{tot} is far above the CMC ($\simeq [S_1^e]$), $\bar{n}S_{\text{mic}} \gg [S_1^e]$. By introducing these conditions into (4.124) the final result is

$$1/\tau_2 \simeq (\bar{n}^2/R[S_1^e]) \times \{1/[1 + \sigma^2 S_{\text{mic}}/(\bar{n}[S_1^e])]\} \tag{4.128}$$

The variation of τ_2 with total surfactant concentration leads to an estimation of the parameters in (4.128).

Turning now to the fast relaxation process, we can find a solution if we make the following assumptions[207,219]:

1. $\vec{k}_n$ and $\overleftarrow{k}_n$ are independent of n over the proper micellar range ($\vec{k}_n = \vec{k}$ and $\overleftarrow{k}_n = \overleftarrow{k}$).
2. The size distribution of micelles obeys the Gaussian distribution

$$[S_n^e] = [S_{\bar{n}}] \exp[-(n - \bar{n})^2/2\sigma^2] \tag{4.129}$$

3. The width of the micellar distribution (σ) is broad enough for n to be treated as a continuous variable, and ξ is also the case.
4. The total number of micelles remains constant during the rapid relaxation period, although the size distribution of micelles changes.

From (4.108) and (4.129) we have

$$\overleftarrow{k}/\vec{k}[S_1^e] = [S_{n-1}^e]/[S_n^e] = \exp(-1/2\sigma^2) \times \exp[(n - \bar{n})/\sigma^2] \tag{4.130}$$

On the other hand, from the above assumptions, we obtain the following differential equation from (4.109):

$$[S_n^e] \times \partial\xi(n, t)/\partial t = (\partial/\partial n)\overleftarrow{k}[S_n^e]\{\partial\xi(n, t)/\partial n - \xi_1[1 + \xi(n, t)]\} \tag{4.131}$$

The relative deviation ξ_n in the micellar region is expanded in the form of L'Hermite polynomials with time-dependent coefficient $C_n(t)$ as

$$\xi(z, t) = \sum_{n=0}^{\infty} C_n(t)H_n(z) \qquad (4.132)$$

$$z = (n - \bar{n})/\sqrt{2}\sigma \qquad (4.133)$$

After insertion of (4.129) and (4.132) into (4.131) and some mathematical manipulation, we have the following equation on the time dependence of monomer concentration $C_1(t)$:

$$dC_1(t)/dt + [\bar{k}/\sigma^2 + \bar{k}a(1 + C_0)/\bar{n}]C_1(t) = -\bar{k}aC_0(1 + C_0)/\sqrt{2}\sigma \qquad (4.134)$$

where

$$a = (S_{tot} - [S_1^e])/[S_1^e] = S_{mic}/[S_1^e] \qquad (4.135)$$

and C_0 is the integration constant. Then, the solution of (4.134) becomes

$$C_1(t) = (C_1(0) + B) \exp(-t/\tau_1) - B \qquad (4.136)$$

where

$$B = [a\sigma C_0(1 + C_0)/\sqrt{2}]/[1 + a\sigma^2(1 + C_0)/\bar{n}] \qquad (4.137)$$

and

$$1/\tau_1 = \bar{k}/\sigma^2 + \bar{k}a(1 + C_0)/\bar{n} \qquad (4.138)$$

The variable τ_1 is the relaxation time for the fast process due to an exchange of surfactant monomer molecules between micelles. Table 4.3 gives values of τ_1 and τ_2 for sodium tetradecyl sulfate micelles at different temperatures and total concentrations.[208] In general, τ_1 and τ_2 decrease with total surfactant concentration and with temperature. In addition, both are found to increase with an increase in the alkyl chain length of homologous surfactants (Fig. 4.16).[208] Variation of τ_2 with surfactant concentration at different ionic strengths leads to the conclusion that at concentrations only slightly above the CMC, both the aggregation number and the rate constant

Table 4.3. Relaxation Times τ_1 (μs) and τ_2 (ms) for Sodium Tetradecyl Sulfate at Different Temperatures and Total Concentrations[a]

$10^3 S_{tot}$, M	15°C τ_1	15°C τ_2	20°C τ_1	20°C τ_2	25°C τ_1	25°C τ_2	30°C τ_1	30°C τ_2	35°C τ_1	35°C τ_2	40°C τ_1	40°C τ_2	45°C τ_1	45°C τ_2
2.1				123	320	41	245	19	155	7		3.5		
2.2				87	290	33	220	13.5	148	5.6		3.3		
2.3		320		95	270	32	203	12	133	5.1	107	2.6		
2.4		298		85	210	32	175	12	131	4.8	108	2.5	87	
2.5		256		90		34		12		4.9		2.4		
2.6				65	183	26	133	10	100	4.2	85	1.9	73	
2.8		235		70	125	34	88	12	73	5.1	60	2	60	
3.0		250		68		29		11		4.3		2		
3.2		258		55		33		11						
3.4		212		62		28								
3.6														
4.0					71		57		43		36		26	
5.0					50		38		29		24		19	
7.5					30		20		14		10		7	
10					17		12		9		6			

[a] Reproduced with permission of the American Chemical Society.

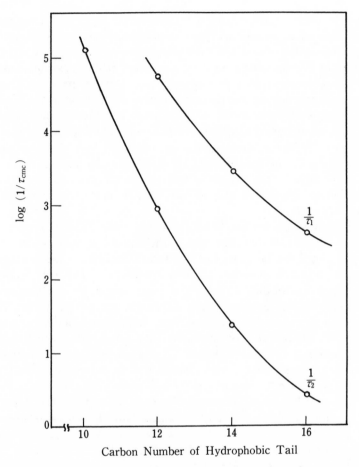

Figure 4.16. Plot of log $(1/\tau_1)_{CMC}$ and log $(1/\tau_2)_{CMC}$ versus the number of carbon atoms n_c in the hydrophobic tail for several sodium alkyl sulfates at 25°C.[208] (Reproduced with permission of the American Chemical Society.)

depend on the surfactant concentration.[220,221] Furthermore, the two relaxation processes have been found to depend on the nature of the counterions,[58] ionic strength,[220] additives or dyes,[222] salts,[190,222] alcohols,[223] and the type of surfactant.[224,225]

4.8. Temperature and Pressure Effects on Micelles

Micelle formation takes place by the aggregation of monomeric surfactant molecules dispersed in a solvent. Aggregation is opposed by both an increase in electrostatic energy (for ionic surfactants) and a decrease in entropy due to aggregation. These unfavorable conditions suggest that micellization is associated with an energy decrease resulting from the condensation of hydrophobic groups (hydrocarbon or fluorocarbon chains) of surfactant molecules into a micellar aggregate. What is the mechanism that causes the energy decrease on the condensation of alkyl chains as a liquidlike hydrocarbon core from dispersed monomers in aqueous medium? Generally, a spontaneous condensation of dispersed solute species in a solution by the van der Waals interaction is accompanied by a decrease in both entropy ($\Delta S < 0$) and free energy ($\Delta G < 0$). However, the entropy change is always positive for the transfer of hydrocarbon from an aqueous environment at infinite dilution to a liquid hydrocarbon phase, for example during micelle formation.

An explanation for this increase involves the peculiar properties of water as a solvent. Water molecules in the liquid state have a structure of hydrogen bonds similar to that of ice, and the cavities in the structure are large enough to accommodate a hydrocarbon chain. The water molecules display equilibria for the formation and destruction of hydrogen bonds with a lifetime of 10^{-12} s, and movement of free water molecules takes place by stepwise jumps through the cavities. Thus, occupation of a cavity by hydrophobic solute hinders the movement of free water molecules, which therefore remain stationary for longer periods. In other words, the water molecules surrounding a hydrophobic solute become more ordered than bulk water molecules and have lower entropy.

Frank and Evans introduced the idea that water molecules form "icebergs" around nonpolar solutes.[226] Nemethy and Scheraga, on the other hand, used the term "increased ice-likeness" to characterize the entropy change.[120] Ben-Naim also suggested a shift into the direction of the "better order" form of water molecules upon introduction of a nobel gas.[227] According to all of these concepts, the water molecules become more ordered around the hydrophobic solute, with an increase in hydrogen bonding in this region. This model can account for the negative enthalpy and entropy of solution. In other words, the favorable free energy for transfer of a nonpolar solute from an aqueous to a hydrophobic environment arises from a large positive entropy associated with the disordering of water molecules in the vicinity of nonpolar solutes. Such interaction between a nonpolar solute and water molecules is termed the *hydrophobic interaction* or *hydrophobic hydration*, and the condensation of the nonpolar solutes by the hydrophobic interaction is conventionally called the *hydrophobic bond*[228] (discussed in more detail by Ben-Naim[229]).

By analogy to the above discussion, the driving force for micellization results from the transfer of nonpolar surfactant chains from an ordered aqueous environment to the hydrocarbonlike environment of the micelle interior, even though a large negative entropy would otherwise have been expected for the transfer of surfactant molecules and counterions from aqueous solution to the confines of a small micelle. With increasing temperature, however, the unique ordered properties of water diminish and water becomes a more normal polar fluid. Consequently, the free energy for transferring a methylene group from the aqueous phase to the micelles has a minimum around 100°C.[230] Conversely, the ordering of water molecules has been reported mostly to disappear around 160°C.[231]

The hydrophilic groups of surfactants make the otherwise hydrophobic molecule somewhat soluble, and at the same time function as a deterrent to micelle formation, keeping the micelle size within a certain range. This occurs because the hydrophilic groups have a great affinity for water molecules by an ion-induced dipole interaction for ionic surfactants and by hydrogen bonding for nonionic surfactants. The nearest-neighbor water molecules of an ionic group come to lack hydrogen bonds under the influence of the strong electric field and become molecules of hydration. The water molecules between these nearest neighbors and the outer bulk water are, therefore, the least restricted, because of the partial destruction of hydrogen bonds, induced by the central ion.

Numerous changes in CMC with temperature are known.[2] Very few of these take place above 100 °C, however,[230,232-234] and studies of the effects of temperature on micellization have mostly been limited to lower temperatures. Figure 4-17 shows an example of CMC change over a wide temperature range.[234] The CMC of ionic surfactants generally passes through a minimum at T_m with increasing temperature. The apparent entropy change of micellization (which is based only on a CMC change with temperature) decreases from a positive value to a negative one at T_m with increasing temperature. The positive entropy change substantiates the idea that the hydrophobic bond supplies a motive force. The negative entropy change, on the other hand, indicates that with increasing temperature the ordinary condensation effect becomes stronger than hydrophobic effects, even though the entropy increase due to the latter still continues above 100°C.

Detailed examination of m and n values over a wide temperature range have made it clear that both variables change with temperature. These changes should be taken into account when considering changes in thermodynamic variables of micelle formation. The micelle aggregation number of ionic surfactants decreases almost linearly with increasing temperature.[221,235-237] The value of m/n decreases as the average aggregation number decreases with increasing temperature.[230,232] This factor makes ΔH_m more positive, which is the right direction of change for explaining another

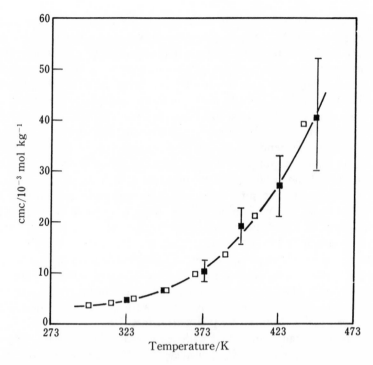

Figure 4.17. Plots of CMC versus temperature of tetradecyltrimethylammonium bromide.[234] ■, enthalpy measurement; □, conductance measurement.[232]

discrepancy noted between a positive ΔH_m obtained from calorimetry and a negative one obtained by using the two-phase theory.[238]

The CMC values for nonionic surfactants are less by one or two orders of magnitude than those of ionic surfactants with the same hydrophobic group, and decrease with rising temperature.[2,237] Specifically, the changes in entropy and enthalpy are positive, and decrease with temperature while remaining positive. The micellar aggregation number of nonionic surfactants increases fairly rapidly with temperature.[237,239,240] As to the aggregation number, one problem remains unsolved: whether the apparent increase of the aggregation number is based on an actual increase in the aggregation number or on an altered spatial arrangement due to increasing interaction among micelles of smaller size as temperature approaches the cloud point.[241,242] The *cloud point* is the temperature at which a nonionic surfactant solution separates into two phases on heating. It depends on the kind of surfactant and on its concentration. This separation is caused by the destruction of hydrogen bonds between water molecules and hydrophillic groups by the thermal turbulence that accompanies rising temperature.

Since the change in CMC with pressure was measured by Tuddenham and Alexander,[112] many reports have appeared on the effect of pressure on micelle formation for ionic[113,115,194,243,244] and nonionic surfactants.[114,245,246] The pressure dependence of the CMC of ionic surfactants is shown in Fig. 4.18 for homologous surfactants, where the CMC values pass through a maximum at around 1000 atm. The partial molar volume change in micelle formation changes from a positive to a negative value as this pressure is exceeded. This change is due to the greater compressibility of surfactant molecules in the micellar state, as is made clear from the following analysis of the data from Fig. 4.18. The partial molar volume change (ΔV_m) obeys a linear relationship against the carbon number (n_c) of the surfactant molecule[113]:

$$\Delta \bar{V}_m = \Delta \bar{V}_{ion} + \Delta \bar{V}_{CH_2} n_c \tag{4.139}$$

where ΔV_{ion} and ΔV_{CH_2} are the contributions of an ionic head group and a methylene group to the partial molar volume change in micelle formation, respectively. The values thus obtained for sodium alkyl sulfates are given in Table 4.4, together with those of SDS. The values of $\Delta \bar{V}_{CH_2}$ change from positive to negative with increasing pressure, which is interpreted as meaning that the compressibility of the micellar surfactants is greater than that of the intermicellar monomer.[247] This result also indicates that hydrophobic bond formation accompanies a positive volume change at lower pressure, whereas at higher pressure it has a negative value.

Such a sign change of $\Delta \bar{V}_m$ is not observed for nonionic surfactants of the polyoxethylene alkyl ether type, because the partial molar volume of the ethylene oxide group in the micellar state increases slightly with pressure, thereby resisting the compression.[248] That is, the CMC value increases monotonously and then levels off with increasing pressure. At the same time, the micellar aggregation number decreases monotonously with rising pressure for nonionic surfactants,[245,246] although the number for ionic surfactants passes through a minimum at around 1000 atm.[248]

The monomeric aqueous solubility of surfactants also depends on pressure, and decreases with increasing pressure. This effect is opposite to that of temperature. In other words, with decreasing pressure the monomeric solubility increases up to the CMC, at which micellization is accompanied by a rapid solubility increase (Fig. 4.19). The pressure at which a solubility-pressure cruve intercepts the CMC-pressure curve is the *critical solution pressure* P_c for micelle formation, which corresponds to the conventional Krafft point for temperature (see Chapter 6). The presence of P_c is observed

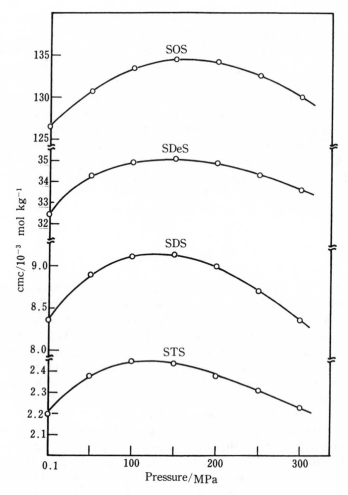

Figure 4.18. Pressure dependence of the CMC for sodium alkyl sulfate at 35°C.[113] SOS, sodium octyl sulfate; SDeS, sodium decyl sulfate; SDS, Sodium dodecyl sulfate; STS, sodium tetradecyl sulfate. (Reproduced with permission of Academic Press.)

for both ionic[249] and nonionic surfactants.[114] Thus, we have a three-dimensional phase diagram with temperature, pressure, and concentration axes (Fig. 4.20).[250]

The size and shape of micelles are very important parameters. However, an introduction to their theoretical background is beyond the scope of this monograph. Those who are interested in more detailed discussion should refer to books on the subject of light scattering.[251-253]

Table 4.4. Contribution of Hydrocarbon Tail and Ionic Head to Partial
Molal Volume Change on Micelle Formation[a]

Pressure (atm)	$\Delta \bar{V}_{CH_2}$ (cm$^3 \cdot$ mol^{-1})	$\Delta \bar{V}_{ion}$ (cm$^3 \cdot$ mol^{-1})	$\Delta \bar{V}_m$ ($n_c = 12$) (cm$^3 \cdot$ mol^{-1})
1	1.24	−4.74	10.1
500	0.47	−2.54	3.1
1000	0.10	−0.30	0.9
1500	−0.236	2.00	−0.8
2000	−0.394	3.24	−1.5
3000	−0.585	3.53	−3.5

[a] Reproduced with permission of Academic Press.[113]

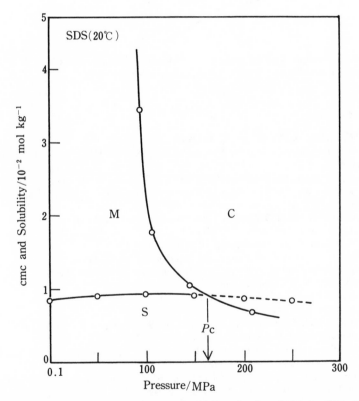

Figure 4.19. Effect of pressure on the CMC and solubility of sodium dodecyl sulfate at 20°C.[249]
M, micellar solution; S, monomer solution; C, hydrated solid; P_c, critical solution pressure.
(Reproduced with permission of Academic Press.)

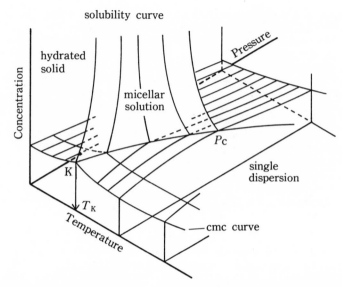

solubility curve

Figure 4.20. Three-dimensional phase diagram of ionic surfactant/water system.[250] T_k, Krafft point; P_c, critical solution pressure; K—P_c, critical solubility curve.

References

1. W. C. Preston, *J. Phys. Chem.* **52**, 84 (1948).
2. P. Mukerjee and K. J. Mysels, *Natl. Stand. Ref. Data Ser.* (U.S. Natl. Bur. Stand.) No. 36 (1971).
3. G. C. Kresheck, in: *Water: A Comprehensive Treatise*, Vol. 4, *Aqueous Solutions of Amphiphiles and Macromolecules* (F. Franks, ed.), pp. 96–98, Plenum Press, New York (1975).
4. R. C. Murray and G. S. Hartley, *Trans. Faraday Soc.* 31, 183 (1935).
5. M. L. Corrin, *J. Colloid Sci.* 3, 333 (1948).
6. J. N. Phillips, *Trans. Faraday Soc.* 51, 561 (1955).
7. E. Matijevic and B. A. Pethica, *Trans. Faraday Soc.* 54, 587 (1958).
8. P. H. Elworthy and K. J. Mysels, *J. Colloid Interface Sci.* 21, 331 (1966).
9. P. Mukerjee, *Adv. Colloid Interface Sci.* 1, 241 (1967).
10. G. Stainsby and A. E. Alexander, *Trans. Faraday Soc.* 46, 587 (1950).
11. K. Shinoda, *Bull. Chem. Soc. Jpn.* 26, 101 (1953).
12. E. Hutchinson, A. Inada, and L. G. Bailey, *Z. Phys. Chem.* (N.F.) 5, 344 (1955).
13. K. Shinoda and E. Hutchinson, *J. Phys. Chem.* 66, 577 (1962).
14. Y. Moroi, N. Nishikido, H. Uehara, and R. Matuura, *J. Colloid Interface Sci.* 50, 254 (1975).
15. P. Mukerjee, *J. Phys. Chem.* 76, 565 (1972).
16. P. Mukerjee, in: *Micellization, Solubilization, and Microemulsions* (K. L. Mittal, ed.), Vol. 1, pp. 171–194, Plenum Press, New York (1977).
17. C. Tanford, *J. Phys. Chem.* 78, 2469 (1974).
18. E. Ruckenstein and R. Nagarajan, *J. Phys. Chem.* 79, 2622 (1975).
19. A. Ben-Naim and F. H. Stillinger, *J. Phys. Chem.* 84, 2872 (1980).
20. E. Ruckenstein and R. Nagarajan, *J. Phys. Chem.* 85, 3010 (1981).

21. E. Hutchinson, *Z. Phys. Chem.* (N.F.) *21*, 38 (1959).
22. C. Botre, V. L. Crescenzi, and A. Mele, *J. Phys. Chem. 63*, 650 (1950).
23. T. Sasaki, M. Hattori, J. Sasaki, and K. Nukina, *Bull. Chem. Soc. Jpn. 48*, 1397 (1975).
24. A. Yamauchi, T. Kunisaki, T. Minematsu, Y. Tomokiyo, T. Yamaguchi, and H. Kimizuka, *Bull. Chem. Soc. Jpn. 51*, 2791 (1978).
25. S. G. Cutler, P. Meares, and D. G. Hall, *J. Chem. Soc. Faraday Trans. 1, 74*, 1758 (1978).
26. K. M. Kale, E. L. Cussler, and D. F. Evans, *J. Phys. Chem. 84*, 593 (1980).
27. K. Shinoda, in: *Colloidal Surfactants*, Chapter 1, Academic Press, New York (1963).
28. F. M. Fowkes, *J. Phys. Chem. 66*, 1843 (1962).
29. J. T. Davies and E. K. Rideal, *Interfacial Phenomena*, p. 201, Academic Press, New York (1963).
30. K. J. Mysels, P. Mukerjee, and M. A. Hamdiyyah, *J. Phys. Chem. 67*, 1943 (1963).
31. K. J. Mysels, *Langmuir 2*, 423 (1986).
32. T. L. Hill, *Thermodynamics of Small Systems*, Vols. 1 and 2, Benjamin, New York (1963, 1964).
33. E. Vikingstad, *J. Colloid Interface Sci. 72*, 68 (1979).
34. T. Maeda and I. Satake, *Bull. Chem. Soc. Jpn. 57*, 2396 (1979).
35. I. Satake, T. Tahara, and R. Matuura, *Bull. Chem. Soc. Jpn. 42*, 319 (1969).
36. Y. Moroi, R. Sugii, and R. Matuura, *J. Colloid Interface Sci. 98*, 184 (1984).
37. M. Hato and K. Shinoda, *Bull. Chem. Soc. Jpn. 46*, 3889 (1973).
38. Y. Moroi, R. Sugii, C. Akine, and R. Matuura, *J. Colloid Interface Sci. 108*, 180 (1985).
39. Y. Moroi, R. Matuura, T. Kuwamura, and S. Inokuma, *J. Colloid Interface Sci. 113*, 225 (1986).
40. J. W. McBain, *Trans. Faraday Soc. 9*, 99 (1913).
41. G. S. Hartle, *Aqueous Solutions of Paraffin-Chain Salts*, Hermann, Paris (1936).
42. J. W. McBain, in: *Colloid Chemistry, Theoretical and Applied* (J. Alexander, ed.), Reinhold, New York (1944).
43. R. H. Mattoon, R. S. Stearns, and W. D. Harkins, *J. Phys. Chem. 15*, 209 (1947); W. D. Harkins, *J. Phys. Chem. 16*, 156 (1948).
44. P. Debye and E. W. Anacker, *J. Phys. Colloid Chem. 55*, 644 (1951).
45. I. Reich, *J. Phys. Chem. 60*, 257 (1956).
46. J. B. Hayter and J. Penfold, *J. Chem. Soc. Faraday Trans. 1 77*, 1851 (1981).
47. D. W. R. Gruen, *Progr. Colloid Polym. Sci. 70*, 6 (1985); *J. Phys. Chem. 89*, 146 (1985), *89*, 153 (1985).
48. B. Cabane, R. Duplessix, and T. Zemb, *J. Phys. (Paris) 46*, 2161 (1985).
49. W. L. Courchene, *J. Phys. Chem. 68*, 1870 (1964).
50. Y. Iwadare and T. Suzawa, *Nippon Kagaku Zasshi 90*, 1106 (1969).
51. D. Bendedouch, S.-H. Chen. and W. C. Koeler, *J. Phys. Chem. 87*, 2621 (1983).
52. Y. Chevalier and C. Chachaty, *Colloid Polym. Sci. 262*, 489 (1984).
53. T. Kawaguchi, T. Hamanaka, and T. Mitsui, *J. Colloid Interface Sci. 96*, 437 (1983).
54. K. S. Birdi, *Prog. Colloid Polym. Sci. 70*, 23 (1985).
55. D. M. Bloor, J. Gormally, and E. Wyn-Jones, *J. Chem. Soc. Faraday Trans. 1 80*, 1915 (1984).
56. R. E. Stark, M. L. Kasakevich, and J. W. Granger, *J. Phys. Chem. 86*, 335 (1982).
57. P. A. Narayana, A. S. W. Li, and L. Kevan, *J. Phys. Chem. 86*, 3 (1982).
58. W. Baumuller, H. Hoffman, W. Ulbricht, C. Tondre, and R. Zana, *J. Colloid Interface Sci. 64*, 418 (1978).
59. I. D. Robb, *J. Colloid Interface Sci. 37*, 521 (1971).
60. D. Stigter and K. J. Mysels, *J. Phys. Chem. 59*, 45 (1955).
61. D. Stigter, *J. Phys. Chem. 78*, 2480 (1974).
62. Y. Chevalier and C. Chachaty, *J. Phys. Chem. 89*, 875 (1985).

63. K. A. Dill, D. E. Koppel, R. S. Cantor, J. D. Dill, D. Bebdedouch, and S.-H. Chen, *Nature 309*, 42 (1984).
64. J. P. Albrizzo and H. E. Cordes, *J. Colloid Interface Sci. 68*, 292 (1979).
65. B. Halle and G. Carlstrom, *J. Phys. Chem. 85*, 2142 (1981).
66. K. S. Birdi, *Colloid Polym. Sci. 252*, 551 (1974).
67. J. C. Ravey, M. Buzier, and C. Picot, *J. Colloid Interface Sci. 97*, 9 (1984).
68. R. J. Williams, J. N. Phillips, and K. J. Mysels, *Trans. Faraday Soc. 51*, 728 (1955).
69. D. G. Hall and B. A. Pethica, in: *Nonionic Surfactants* (M. J. Schick, ed.), Chapter 16, Dekker, New York (1967).
70. J. N. Israelachvili, D. J. Mitchell, and B. W. Ninham, *J. Chem. Soc. Faraday Trans. 2 72*, 1525 (1976).
71. P. Mukerjee and A. Y. S. Yang, *J. Phys. Chem. 80*, 1388 (1976).
72. Y. Moroi and R. Matuura, *Bull. Chem. Soc. Jpn. 61*, 333 (1988).
73. P. Debye, *J. Phys. Chem. 53*, 1 (1949).
74. H. Lange, *Parfume. Kosmet. 46*, 129 (1965).
75. D. J. Shaw, *Introduction to Colloid and Surface Chemistry*, 2nd ed., p. 65, Butterworths, London (1970).
76. P. Becher, in: *Nonionic Surfactants* (M. J. Schick, ed.), Chapter 15, Dekker, New York (1967).
77. P. Becher, *J. Colloid Sci. 16*, 49 (1961).
78. M. J. Schick, S. M. Atlas, and F. R. Eirich, *J. Phys. Chem. 66*, 1326 (1962).
79. P. Becher, in: Proceedings 4th International Congress on Surface Active Substances, Brussels 1964 (J. T. G. Overbeek, ed.), Vol. 2, p. 621.
80. J. E. Desnoyers, G. Caron, R. DeLisi, D. Roberts, A. Roux, and G. Perron, *J. Phys. Chem. 87*, 1397 (1983).
81. D. C. Poland and H. A. Scheraga, *J. Phys. Chem. 69*, 2431 (1965).
82. Y. Moroi and R. Matuura, *J. Phys. Chem. 89*, 2923 (1985).
83. Y. Moroi, *J. Colloid Interface Sci. 122*, 308 (1988).
84. P. Mukerjee, K. J. Mysels, and C. I. Dulin, *J. Phys. Chem. 62*, 1390 (1958).
85. L. Shedlovsky, C. W. Jakob, and M. B. Epstein, *J. Phys. Chem. 67*, 2975 (1963).
86. D. Eagland and F. Franks, *Trans. Faraday Soc. 61*, 2468 (1965).
87. T. Kawai, J. Umemura, and T. Takenaka, *Colloid Polym. Sci. 262*, 61 (1984).
88. M. J. Vold, *J. Colloid Interface Sci. 116*, 129 (1987).
89. M. L. Corrin and W. D. Harkins, *J. Am. Chem. Soc. 69*, 683 (1947).
90. M. E. Hobbs, *J. Phys. Chem. 55*, 675 (1951).
91. M. J. Schick, *J. Phys. Chem. 68*, 3585 (1964).
92. M. F. Emerson and A. Holtzer, *J. Phys. Chem. 71*, 1898 (1967).
93. E. J. R. Sudholter and J. B. F. N. Engberts, *J. Phys. Chem. 83*, 1854 (1979).
94. H. N. Singh, S. Swarup, and S. M. Saleem, *J. Colloid Interface Sci. 68*, 128 (1979).
95. Y. Moroi and Y. Sakamoto, *J. Phys. Chem. 92*, 5189 (1988).
96. C. A. J. Hoeve and G. C. Benson, *J. Phys. Chem. 61*, 1149 (1957).
97. T. Nakagawa, K. Juriyama, and K. Tohri, *Nippon Kagaku Zasshi 78*, 1568 (1957).
98. D. C. Poland and H. A. Scheraga, *J. Colloid Interface Sci. 21*, 273 (1966).
99. H. S. Chung and I. J. Heilweil, *J. Phys. Chem. 74*, 488 (1970).
100. W. E. McMullen III, W. M. Gelbart, and A. Ben-Shaul, *J. Phys. Chem. 88*, 6649 (1984).
101. J. M. Corkhill, J. F. Goodman, and S. P. Harrold, *Trans. Faraday Soc. 60*, 202 (1984).
102. J. M. Corkhill, J. F. Goodman, T. Walker, and J. Wyer, *Proc. R. Soc. London Ser. A 312*, 243 (1969).
103. P. F. Mihnlieff, *J. Colloid Interface Sci. 33*, 255 (1970).
104. D. G. Hall, *J. Chem. Soc. Faraday Trans. 2 68*, 1439 (1972); *73*, 897 (1977).

105. B. Jonsson and H. Wennerstrom, *J. Colloid Interface Sci.* 80, 482 (1981).
106. A. I. Rusanov, *J. Colloid Interface Sci.* 85, 157 (1982).
107. J. E. Desnoyers, *Pure Appl. Chem.* 54, 1469 (1982).
108. H. Akisada, *J. Colloid Interface Sci.* 97, 105 (1984).
109. T. E. Burchfield and E. M. Wooley, *J. Phys. Chem.* 88, 2149 (1984); 88, 2155 (1984); 89, 714 (1985).
110. D. G. Hall and R. W. Huddleston, *Colloids Surfaces* 13, 209 (1985).
111. Y. Moroi, R. Matuura, T. Kuwamura, and S. Inokuma, *Colloid Polym. Sci.* 266, 374 (1988).
112. R. F. Tuddenham and A. E. Alexander, *J. Phys. Chem.* 66, 1839 (1962).
113. S. Kaneshina, M. Tanaka, T. Tomida, and R. Matuura, *J. Colloid Interface Sci.* 48, 450 (1974).
114. N. Nishikido, N. Yoshimura, M. Tanaka, and S. Kaneshina, *J. Colloid Interface Sci.* 78, 338 (1980).
115. M. Yamanaka, M. Aratono, K. Motomura, and R. Matuura, *Colloid Polym. Sci.* 262, 338 (1984).
116. K. Shinoda and T. Soda, *J. Phys. Chem.* 67, 2072 (1963).
117. J. M. Corkhill, J. F. Goodman, and T. Walker, *Trans. Faraday Soc.* 63, 768 (1967).
118. G. M. Musbally, G. Perron, and J. E. Desnoyers, *J. Colloid Interface Sci.* 48, 494 (1974).
119. A. H. Roux, D. Hetu, G. Perron, and J. E. Desnoyers, *J. Solution Chem.* 13, 1 (1984).
120. G. Nemethy and H. A. Scheraga, *J. Chem. Phys.* 36, 3401 (1962).
121. E. Matijevic and B. A. Pethica, *Trans. Faraday Soc.* 54, 587 (1958).
122. P. H. Elworthy and A. T. Florence, *J. Pharm. Pharmacol.* 15, 851 (1963).
123. F. Tokiwa and K. Ohki, *Kolloid Z. Z. Polym.* 223, 138 (1967).
124. B. W. Barry and D. I. D. E. Eini, *J. Colloid Interface Sci.* 54, 339 (1976).
125. E. D. Goddard and G. C. Benson, *Trans. Faraday Soc.* 52, 409 (1956).
126. P. White and G. C. Benson, *Trans. Faraday Soc.* 55, 1025 (1959).
127. G. Pilcher, M. N. Jones, L. Espada, and H. A. Skinner, *J. Chem. Thermodyn.* 1, 381 (1969); 2, 1 (1970); 2, 333 (1970); 3, 801 (1971).
128. M. N. Jones and J. Piercy, *Kolloid Z. Z. Polym.* 251, 343 (1973).
129. G. C. Kresheck and W. A. Hargraves, *J. Colloid Interface Sci.* 48, 481 (1974).
130. J. H. Clint and T. Walker, *J. Chem. Soc. Faraday Trans. 1* 76, 946 (1975).
131. J. L. Woodhead, J. A. Lewis, G. N. Malcolm, and I. D. Watson, *J. Colloid Interface Sci.* 79, 454 (1981).
132. P. White and G. C. Benson, *J. Colloid Sci.* 13, 584 (1958).
133. M. J. Schick, *J. Phys. Chem.* 67, 1796 (1963).
134. S. Paredes, M. Tribout, J. Ferreira, and J. Leonis, *Colloid Polym. Sci.* 254, 637 (1976).
135. K. S. Birdi, *Colloid Polym. Sci.* 261, 45 (1983).
136. E. D. Goddard, C. A. J. Hoeve, and G. C. Benson, *J. Phys. Chem.* 61, 593 (1957).
137. L. Benjamin, *J. Phys. Chem.* 68, 3575 (1964).
138. J. M. Corkill, J. F. Goodman, S. P. Harrold, and J. R. Tate, *Trans. Faraday Soc.* 62, 994 (1966).
139. D. J. Mitchell and B. W. Ninham, *J. Chem. Soc. Faraday Trans. 2*, 77, 601 (1981).
140. D. Stigter, *J. Phys. Chem.* 68, 3603 (1964).
141. D. Stigter, *J. Phys. Chem.* 79, 1008 (1975).
142. D. Stigter, *J. Phys. Chem.* 79, 1015 (1975).
143. G. Gunnarsson, B. Jonsson, and H. Wennerstrom, *J. Phys. Chem.* 84, 3114 (1980).
144. P. Linse, G. Gunnarsson, and B. Jonsson, *J. Phys. Chem.* 86, 413 (1982).
145. D. F. Evans and B. W. Ninham, *J. Phys. Chem.* 87, 5025 (1983).
146. K. Shirahama, *Bull. Chem. Soc. Jpn.* 47, 3165 (1974).
147. T. Kaneko, *Bull. Chem. Soc. Jpn.* 59, 1290 (1986).

148. A. L. Loeb, J. T. G. Overbeek, and P. H. Wiersema, *The Electric Double Layer around a Spherical Colloid Particle,* MIT Press, Cambridge, Mass. (1961).
149. F. Oosawa, *Polyelectrolytes,* Dekker, New York (1971).
150. N. Kamenka, B. Lindman, K. Fontell, M. Chorro, and B. Brun, *C.R. Acad. Sci. 284,* 403 (1977).
151. E. W. Anacker and H. M. Ghose, *J. Am. Chem. Soc. 90,* 3161 (1968).
152. D. Bartet, C. Gamboa, and L. Sepulveda, *J. Phys. Chem. 84,* 272 (1980).
153. P. Mukerjee, K. J. Mysels, and P. Kapauan, *J. Phys. Chem. 71,* 4166 (1967).
154. E. W. Anacker and A. L. Underwood, *J. Phys. Chem. 85,* 2463 (1981).
155. A. L. Underwood and E. W. Anacker, *J. Colloid Interface Sci. 100,* 128 (1984).
156. A. L. Underwood and E. W. Anacker, *J. Phys. Chem. 88,* 2390 (1984).
157. M. Almgren and S. Swarup, *J. Phys. Chem. 87,* 876 (1983).
158. E. Lissi, E. Abuin, I. M. Cuccovia, and H. Chaimovich, *J. Colloid Interface Sci. 112,* 513 (1986).
159. R. Zana, *J. Colloid Interface Sci. 78,* 330 (1980).
160. J. A. Beunen and E. Ruckenstein, *J. Colloid Interface Sci. 96,* 469 (1983).
161. J. W. Larsen and L. B. Tepley, *J. Colloid Interface Sci. 49,* 113 (1974).
162. M. Koshinuma, *Bull. Chem. Soc. Jpn. 52,* 1790 (1979).
163. S. Miyamoto, *Bull. Chem. Soc. Jpn 33,* 371 (1960); *33,* 375 (1960).
164. I. Satake, I. Iwamatsu, S. Hosokawa, and R. Matuura, *Bull. Chem. Soc. Jpn. 36,* 204 (1963).
165. I. Satake and R. Matuura, *Bull. Chem. Soc. Jpn 36,* 813 (1963).
166. Y. Moroi, K. Motomura, and R. Matuura, *J. Colloid Interface Sci. 46,* 111 (1974).
167. Y. Moroi, N. Nishikido, and R. Matuura, *J. Colloid Interface Sci. 50,* 344 (1975).
168. A. Yamauchi, H. Tokunaga, S. Matsuno, and H. Kimizuka, *Nippon Kagaku Kaishi 1980,* 388.
169. H. Gustavsson and B. Lindman, *J. Am. Chem. Soc. 97,* 3923 (1975).
170. M. Jansson and P. Stilbs, *J. Phys. Chem. 91,* 113 (1987).
171. C. Tanford, *Proc. Natl. Acad. Sci. USA 71,* 1811 (1974).
172. G. Kegeles, *J. Phys. Chem. 83,* 1728 (1979).
173. H. D. Young, *Statistical Treatment of Experimental Data,* McGraw-Hill, New York (1962).
174. H. V. Tartar, *J. Phys. Chem. 59,* 1195 (1955).
175. C. Tanford, *J. Phys. Chem. 76,* 3020 (1972).
176. J. E. Leibner and J. Jacobus, *J. Phys. Chem. 81,* 130 (1977).
177. M. J. Vold, *Langmuir 1,* 501 (1985).
178. M. Miura and M. Kodama, *Bull. Chem. Soc. Jpn. 45,* 428 (1972); *45,* 2265 (1972); *45,* 2953 (1972).
179. F. Quirion and J. E. Desnoyers, *J. Colloid Interface Sci. 112,* 565 (1986).
180. A. Malliaris, J. Lang, and R. Zana, *J. Colloid Interface Sci. 110,* 237 (1986).
181. L. M. Kushner, W. D. Hubbard, and R. A. Parker, *J. Res. Natl. Bur. Stand. 59,* 113 (1957).
182. J. P. Kratohvil, *J. Colloid Interface Sci. 75,* 271 (1980).
183. K. Kalyanasundaram, M. Gratzel, and J. K. Thomas, *J. Am. Chem. Soc. 97,* 3915 (1975).
184. N. A. Mazer, G. B. Benedek, and M. C. Carey, *J. Phys. Chem. 80,* 1075 (1976).
185. S. Ikeda, S. Ozeki, and M. Tsunoda, *J. Colloid Interface Sci. 73,* 27 (1980).
186. S. Ikeda, S. Ozeki, and S. Hayashi, *Biophys. Chem. 11,* 417 (1980).
187. S. Hayashi and S. Ikeda, *J. Phys. Chem. 84,* 744 (1980).
188. S. Ikeda, S. Hayashi, and T. Imae, *J. Phys. Chem. 85,* 106 (1981).
189. S. Ozeki and S. Ikeda, *J. Colloid Interface Sci. 87,* 424 (1982).
190. E. Lessner and J. Frahm, *J. Phys. Chem. 86,* 3032 (1982).
191. T. Imae and S. Ikeda, *J. Phys. Chem. 90,* 5216 (1986).
192. G. Lindblom, B. Lindman, and L. Mandell, *J. Colloid Interface Sci. 42,* 400 (1973).
193. E. Hirsch, S. Candau, and R. Zana, *J. Colloid Interface Sci. 97,* 318 (1984).

194. E. Ljosland, A. M. Blokhus, K. Veggeland, S. Backlund, and H. Hoiland, *Prog. Colloid Polym. Sci.* 74, 34 (1985).
195. I. Vikholm, G. Douheret, S. Backlund, and H. Hoiland, *J. Colloid Interface Sci.* 116, 582 (1987).
196. P. J. Missel, N. A. Mazer, G. B. Benedek, C. Y. Young, and M. C. Carey, *J. Phys. Chem.* 84, 1044 (1980).
197. P. J. Missel, N. A. Mazer, G. B. Benedek, and M. C. Carey, *J. Phys. Chem.* 87, 1264 (1983).
198. S. Ikeda, *J. Phys. Chem.* 88, 2144 (1984).
199. G. M. Thurston, D. Blackschtein, M. R. Fish, and G. B. Benedek, *J. Chem. Phys.* 84, 4558 (1986).
200. A. Ben-Shaul, D. H. Rorman, G. V. Hartland, and W. M. Gelbart, *J. Phys. Chem.* 90, 5277 (1986).
201. T. L. Hill, *An Introduction to Statistical Thermodynamics*, Addison–Wesley, Reading, Mass. (1960).
202. J. F. McKellar and J. H. Andreae, *Nature* 195, 778 (1962); 195, 865 (1962).
203. G. C. Kresheck, E. Hamori, G. Davenport, and H. A. Scheraga, *J. Am. Chem. Soc.* 88, 246 (1966).
204. E. Graber, J. Lang, and R. Zana, *Kolloid Z. Z. Polym.* 238, 470 (1970).
205. K. Takeda and T. Yasunaga, *J. Colloid Interface Sci.* 40, 127 (1972).
206. J. Rassing, P. J. Sams, and E. Wyn-Jones, *J. Chem. Soc. Faraday Trans. 2* 70, 1247 (1974).
207. E. A. G. Aniansson and S. N. Wall, *J. Phys. Chem.* 78, 1024 (1974); 79, 857 (1975).
208. E. A. G. Aniansson, S. N. Wall, M. Almgren, H. Hoffmann, I. Kielmann, W. Ulbricht, R. Zana, J. Lang, and C. Tondre, *J. Phys. Chem.* 80, 905 (1976).
209. T. Nakagawa, *Colloid Polym. Sci.* 252, 56 (1974).
210. M. J. Jaycock and R. H. Ottewill, Proc. 4th Int. Congr. Surface Active Substances, 1964, Section B, paper 8.
211. T. Yasunaga, K. Takeda, and S. Harada, *J. Colloid Interface Sci.* 42, 457 (1973).
212. P. F. Mijnkieff and R. Ditmarsch, *Nature* 208, 889 (1965).
213. B. C. Bennion and E. M. Eyring, *J. Colloid Interface Sci.* 32, 286 (1970).
214. T. Yasunaga, H. Oguri, and M. Miura, *J. Colloid Interface Sci.* 23, 352 (1967).
215. E. Graber and R. Zana, *Kolloid Z. Z. Polym.* 238, 479 (1970).
216. T. Nakagawa, H. Inoue, H. Jizomoto, and K. Horiuchi, *Kolloid Z. Z. Polym.* 229, 159 (1969).
217. K. K. Fox, *Trans. Faraday Soc.* 67, 2802 (1971).
218. J. Lang, and E. M. Eyring, *J. Polym. Sci. A-2* 10, 89 (1972).
219. S. N. Wall and G. E. A. Aniansson, *J. Phys. Chem.* 84, 727 (1980).
220. J. Lang, C. Tondre, R. Zana, R. Bauer, H. Hoffmann, and W. Ulbricht, *J. Phys. Chem.* 79, 276 (1975).
221. M. Grubic, R. Strey, and M. Teubner, *J. Colloid Interface Sci.* 80, 453 (1981).
222. C. Tondre, J. Lang, and R. Zana, *J. Colloid Interface Sci.* 52, 372 (1975).
223. J. Lang and R. Zana, *J. Phys. Chem.* 90, 5258 (1986).
224. C. Tondre and R. Zana, *J. Colloid Interface Sci.* 66, 544 (1978).
225. T. Inoue, T. Tashiro, Y. Shibuya, and R. Shimozawa, *J. Colloid Interface Sci.* 73, 105 (1980).
226. H. S. Frank and M. W. Evans, *J. Phys. Chem.* 13, 507 (1945).
227. A. Ben-Naim, *J. Phys. Chem.* 69, 3240 (1965).
228. W. Kauzmann, *Adv. Protein Chem.* 44, 1 (1959).
229. A. Ben-Naim, *Hydrophobic Interactions*, Plenum Press, New York (1980).
230. D. F. Evans, M. Allen, B. W. Ninham, and A. Fouda, *J. Solution Chem.* 13, 87 (1984).
231. K. Shinoda, *J. Phys. Chem.* 81, 1300 (1977).
232. D. F. Evans and P. J. Wightman, *J. Colloid Interface Sci.* 86, 515 (1982).

233. D. G. Archer, H. J. Albert, D. E. White, and R. H. Wood, *J. Colloid Interface Sci.* 100, 68 (1984).
234. D. G. Archer, *J. Solution Chem.* 16, 347 (1987).
235. M. Jones and J. Piercy, *J. Chem. Soc. Faraday Trans. 1* 68, 1839 (1972).
236. Y. Croonen, E. Gelade, M. Van der Zegel, M. Van der Auweraer, H. Vandendriessche, F. C. De Schyryver, and M. Almgren, *J. Phys. Chem.* 87, 1426 (1983).
237. A. Malliaris, J. L. Moigne, J. Sturm, and R. Zana, *J. Phys. Chem.* 89, 2709 (1985).
238. P. Mukerjee, *J. Phys. Chem.* 66, 1375 (1962).
239. M. J. Schick, *Nonionic Surfactants*, Dekker, New York (1967).
240. W. Brown, R. Johnson, P. Stilbs, and B. Lindman, *J. Phys. Chem.* 87, 4548 (1983).
241. M. Zulauf and J. P. Rosenbusch, *J. Phys. Chem.* 87, 856 (1983).
242. D. J. Cebula and R. H. Ottewill, *Colloid Polym. Sci.* 260, 1118 (1982).
243. M. Yamanaka, M. Aratono, and K. Motomura, *Bull. Chem. Sco. Jpn* 59, 2695 (1986).
244. Y. Ikawa, S. Tsuru, Y. Murata, M. Okawauchi, M. Shigematsu, and G. Sugihara, *J. Solution Chem.* 17, 125 (1988).
245. N. Nishikido, M. Shinozaki, G. Sugihara, and M. Tanaka, *J. Colloid Interface Sci.* 82, 352 (1981).
246. M. Okawauchi, M. Shinoazki, Y. Ikawa, and M. Tanaka, *J. Phys. Chem.* 91, 109 (1987).
247. M. Tanaka, S. Kaneshina, K. Shin-no, T. Okajima, and T. Tomida, *J. Colloid Interfact Sci.* 46, 132 (1974).
248. N. Nishikido, M. Shinozaki, G. Sugihara, M. Tanaka, and S. Kaneshina, *J. Colloid Interface Sci.* 74, 474 (1980).
249. M. Tanaka, S. Kaneshina, T. Tomida, K. Noda, and K. Aoki, *J. Colloid Interface Sci.* 44, 525 (1973).
250. M. Tanaka, S. Kaneshina, G. Sugihara, N. Nishikido, and Y. Murata, in: *Solution Behavior of Surfactants* (K. L. Mittal, ed.), Vol. 1, p. 41, Plenum Press, New York (1982).
251. C. Tanford, *Physical Chemistry of Macromolecules*, Chapter 5, Wiley, New York (1961).
252. B. J. Berne and R. Pecora, *Dynamic Light Scattering: With Applications to Chemistry, Biology, and Physics*, Wiley, New York (1978).
253. P. Kratochvil, *Classical Light Scattering from Polymer Solutions*, Czechoslovak Academy of Sciences, Prague (1987).

5

Application of the Thermodynamics of Small Systems to Micellar Solutions

5.1. Introduction

The theoretical treatments of micelle formation have been based mainly on the mass-action and phase-separation models, even though the phase-separation model is inconsistent with the degrees of freedom of the phase rule. In addition, the two models merge asymptotically with increasing micellar aggregation number. As discussed in the preceding chapters, the mass-action model requires knowledge of all of the stepwise association constants from monomer to micelles, a requirement difficult to fulfill. This model therefore has such defects as the assumption that the micelle aggregation number is monodisperse (in spite of its actual polydispersity) or the fact that some numerical values for micellization constants are assumed when estimating the dispersion of micellar size.

   Hill developed the thermodynamics of small systems and also applied it to the aggregation of solutes.[1] This theory serves as a bridge between the mass-action and phase-separation models. Further development has been done by Hall.[2-4] Recently, Tanaka applied the theory to static light scattering data for aqueous solutions of nonionic surfactants, and proved its usefulness.[5] This chapter introduces the fundamental concept of this thermodynamics as a basis for understanding micellar solutions.

5.2. Small Systems Not in Solution

   Let us consider an ensemble of N small systems that are open to C components, where N is large enough for the ensemble to be treated by conventional thermodynamics. The total Gibbs free energy (G_t) of the ensemble is a function of temperature, pressure, the total number of molecules (N_i^t) of component i, and the number of small systems:

$$G_t = G_t(T, P, N_i^t, N) \quad \text{and} \quad N_i^t = \bar{N}_i N \tag{5.1}$$

where $\bar{N}_i$ is the average number of component i per small system. Then we have

$$dG_t = -S_t \, dT + V_t \, dP + \sum_i \mu_i \, dN_i^t + \xi \, dN \tag{5.2}$$

where S_t and V_t are the entropy and volume of the ensemble, respectively. From (5.2) we obtain

$$\mu_i = (\partial G_t / \partial N_i^t)_{T,P,N} \tag{5.3}$$

$$\xi = (\partial G_t / \partial N)_{T,P,N_i^t} \tag{5.4}$$

The term $\xi \, dN$ in (5.2) is called the *subdivision energy*, and ξ is the *subdivision potential*, that is, the energy term necessary to divide small systems by increasing their number when temperature, pressure, and the total number of molecules are kept constant. When the average number of molecules per small system $\bar{N}_i$ increases, the thermodynamic variables approach those of macroscopic systems, and ξ vanishes.

On the other hand, when a closed ensemble of open systems is in equilibrium at constant T and P, the Gibbs free energy of the ensemble should be minimal $(dG_t = 0)$:

$$(\partial G_t / \partial N)_{T,P,N_i^t} = \xi = 0 \tag{5.5}$$

For a stable equilibrium:

$$(\partial^2 G_t / \partial N^2)_{T,P,N_i^t} = (\partial \xi / \partial N)_{T,P,N_i^t} \geq 0 \tag{5.6}$$

The extensive thermodynamic variables are a first-order homogeneous function of $\bar{N}_i^t$ and N, a relation known as Euler's theorem[6]:

$$G_t(T, P, \lambda N_i^t, \lambda N) = \lambda G_t(T, P, N_i^t, N) \tag{5.7}$$

Integrating (5.2) with respect to N_i^t and N under constant T and P leads to

$$G_t = \sum_i \mu_i N_i^t + \xi N \tag{5.8}$$

From (5.2) and (5.8) we now have

$$-S_t\,dT + V_t\,dP - \sum_i N_i^t\,d\mu_i - \mathsf{N}\,d\xi = 0 \tag{5.9}$$

This is the Gibbs–Duhem equation of the ensemble. When the extensive variables per small system are defined by $\bar{X} = X_t/\mathsf{N}$, (5.8) and (5.9) are rewritten as

$$\bar{G} = \sum_i \mu_i \bar{N}_i + \xi \tag{5.10}$$

$$d\xi = -\bar{S}\,dT + \bar{V}\,dP - \sum_i \bar{N}_i\,d\mu_i \tag{5.11}$$

From (5.10) and (5.11) we have

$$d\bar{G} = -\bar{S}\,dT + \bar{V}\,dP + \sum_i \mu_i\,d\bar{N}_i \tag{5.12}$$

Equation (5.12) indicates that macroscopic thermodynamics is applicable to a small system. However, $\bar{G}$ is found not to be a homogeneous function of $\bar{N}_i$, as is clear from (5.10):

$$\bar{G} \neq \sum_i \mu_i \bar{N}_i \quad \text{or} \quad \bar{G}(T, P, \lambda\bar{N}_i) \neq \lambda\bar{G}(T, P, \bar{N}_i) \tag{5.13}$$

5.3. Small Systems in Solution

When small systems exist in a solution, they are open to C components in the solution, and temperature, pressure, and chemical potential μ_i become the environmental variables for the small systems. Here the following mass balances are set up:

$$N_i^t = N_i^s + N_i^m \tag{5.14}$$

and

$$N_i^s = \mathsf{N}\bar{N}_i^s \quad \text{and} \quad N_i^m = \mathsf{N}\bar{N}_i \tag{5.15}$$

where N_i^s and N_i^m are the numbers of molecules of component i in solvent and micelle, respectively. Now the Gibbs free energy of the ensemble becomes a function of temperature, pressure, N_i^s, N_i^m, and the number of small systems N:

$$G_t = G_t(T, P, N_i^s, N_i^m, \mathsf{N}) \tag{5.16}$$

Then we have

$$dG_t = -S_t \, dT + V_t \, dP + \sum_i \mu_i^s \, dN_i^s + \sum_i \mu_i^m \, dN_i^m + \xi \, dN \qquad (5.17)$$

For a closed ensemble ($N_i^t = $ constant), (5.17) becomes

$$dG_t = -S_t \, dT + V_t \, dP + \sum_i N(\mu_i^m - \mu_i^s) \, d\bar{N}_i + \left(\hat{\mu} - \sum_i \bar{N}_i \mu_i^s \right) dN \qquad (5.18)$$

where

$$\hat{\mu} = \sum_i \bar{N}_i \mu_i^m + \xi \qquad (5.19)$$

If the ensemble is at constant T and P, (5.18) becomes

$$dG_t = \sum_i N(\mu_i^m - \mu_i^s) \, d\bar{N}_i + \left(\hat{\mu} - \sum_i \bar{N}_i \mu_i^s \right) dN \qquad (5.20)$$

The conditions for aggregation equilibrium ($dG_t \geq 0$) are therefore

$$\mu_i^s = \mu_i^m (=\mu_i) \quad \text{and} \quad \hat{\mu} = \sum_i \bar{N}_i \mu_i^s (\text{or } \xi = 0) \qquad (5.21)$$

Equation (5.21) is the condition for a closed ensemble at equilibrium. Equations (5.14) and (5.21), together with (5.17), lead to the following equation, which is the same as that for an ordinary macroscopic system:

$$dG_t = -S_t \, dT + V_t \, dP + \sum_i \mu_i \, dN_i^t \qquad (5.2')$$

This is the same as (5.2) for the case where $\xi = 0$. For a closed ensemble at equilibrium, $C + 2$ variables (T, P, and C) determine the state of the entire thermodynamic system, including its size. Because of the intensive properties are independent of the size, they are specified by $C + 1$ independent intensive variables: T, P, and $C - 1$ mole fractions of components. This is easily understood by the Gibbs–Duhem equation from (5.2'); any intensive property (T, P, μ_i) is a function of $C + 1$ intensive variables. The number of degrees of freedom of the system is therefore identical to that of the ususal macroscopic system, that is, $C + 1$ for C components in one phase.

On the other hand, the differential of N_i^m is given by

$$dN_i^m = N \, d\bar{N}_i + \bar{N}_i \, dN \qquad (5.22)$$

From (5.17) and (5.22) we have

$$dG_t = -S_t\, dT + V_t\, dP + \sum_i \mu_i^s\, dN_i^s + \hat{\mu}\, dN + \sum_i N\mu_i^m\, d\bar{N}_i \quad (5.23)$$

where

$$\hat{\mu} = (\partial G_t/\partial N)_{T,P,N_i^s,\bar{N}_i} = \sum_i \bar{N}_i\mu_i^m + \xi \quad (5.24)$$

and $\hat{\mu}$ is the chemical potential of a micelle. Because G_t is the linear homologous function of N_i^s and N at constant temperature, pressure, and micellar composition, G_t is written as

$$G_t = \sum_i \mu_i^s N_i^s + \hat{\mu}N \quad (5.25)$$

From (5.23) and (5.25'),

$$-S_t\, dT + V_t\, dP - \sum_i N_i^s\, d\mu_i^s - N\, d\hat{\mu} + \sum_i N\mu_i^m\, d\bar{N}_i = 0 \quad (5.26)$$

or

$$d\hat{\mu} = -\bar{S}\, dT + \bar{V}\, dP - \sum_i \bar{N}_i^s\, d\mu_i^s + \sum_i \mu_i^m\, d\bar{N}_i \quad (5.27)$$

Let us consider a macroscopic system at T and P composed of $\bar{N}_i^s$ molecules that have chemical μ_i^s, where the system contains no small system (or micelle). The Gibbs–Duhem equation for the system becomes

$$-\bar{S}^s\, dT + \bar{V}^s\, dP - \sum_i \bar{N}_i^s\, d\mu_i^s = 0 \quad (5.28)$$

Subtracting (5.28) from (5.27) yields

$$d\hat{\mu} = -\hat{S}\, dT + \hat{V}\, dP + \sum_i \mu_i^m\, d\bar{N}_i \quad (5.29)$$

and

$$\hat{S} = \bar{S} - \bar{S}^s \quad \text{and} \quad \hat{V} = \bar{V} - \bar{V}^s \quad (5.30)$$

where the thermodynamic variables with superscript $\char"5E$ refer to the intrinsic variables of the micelle. The chemical potential of component i in the micellar state is then given by

$$\mu_i^m = (\partial\hat{\mu}/\partial\bar{N}_i)_{T,P,\bar{N}_j} \quad (5.31)$$

Equations (5.23), (5.29), and (5.31) are basic equations for small systems in solution, where $\hat{\mu}$ is a function of T, P, $\bar{N}_i$, N_i^s, and N, as is clear from (5.23). From (5.19) and (5.29) we have

$$d\xi = -\hat{S}\,dT + \hat{V}\,dP - \sum_i \bar{N}_i\,d\mu_i^m \tag{5.32}$$

On the other hand, $\hat{\mu}$ can be regarded as a function of T, P, $\bar{N}_i$, and C_m through the following relation:

$$C_m = N \Big/ \left(\sum_i N_i^s + N \right) \simeq N/N_o^s \tag{5.33}$$

where C_m is a mole fraction of the small systems in the ensemble, and the component o refers to a solvent. Because N_o^s is much larger than N_i^s or N, the second equality of (5.33) is satisfied. For a dilute solution of small systems where interaction between the systems is negligible, $\hat{\mu}$ can be written as

$$\hat{\mu}(T, P, \bar{N}_i, C_m) = G(T, P, \bar{N}_i) + kT \ln C_m \tag{5.34}$$

where G is the standard chemical potential of a small system, implicitly including the interaction with solvent. From (5.31) and (5.34) we have

$$\mu_i^m = (\partial G/\partial \bar{N}_i)_{T,P,\bar{N}_j} \tag{5.35}$$

because C_m is not a function of $\bar{N}_i$. Furthermore, (5.19) and (5.34) lead to

$$G(T, P, \bar{N}_i) = \sum_i \bar{N}_i\mu_i^m + \xi_m \tag{5.36}$$

where

$$\xi_m = \xi - kT \ln C_m \tag{5.37}$$

or

$$\xi_m = -kT \ln C_m \quad \text{(at equilibrium; } \xi = 0) \tag{5.37'}$$

The differential of G is then given by

$$dG(T, P, \bar{N}_i) = -S\,dT + V\,dP + \sum_i \mu_i^m\,d\bar{N}_i \tag{5.38}$$

From (5.36) and (5.38) there results

$$d\xi_m = -S\,dT + V\,dP - \sum_i \bar{N}_i\,d\mu_i^m \tag{5.39}$$

For a closed ensemble at equilibrium ($\mu_i^s = \mu_i^m = \mu_i$, $\xi = 0$) we have the following fundamental equations:

$$\xi_m = -kT\ln C_m \tag{5.37'}$$

$$G = \sum_i \bar{N}_i\mu_i - kT\ln C_m \tag{5.36'}$$

$$dG = -S\,dT + V\,dP + \sum_i \mu_i\,d\bar{N}_i \tag{5.38'}$$

$$-d(kT\ln C_m) = -S\,dT + V\,dP - \sum_i \bar{N}_i\,d\mu_i \tag{5.39'}$$

$$-k\,d\ln C_m = -(H/T^2)\,dT + (V/T)\,dP - \sum_i \bar{N}_i\,d(\mu_i/T) \tag{5.40}$$

5.4. Size Distribution of Micelles

Micelles are not monodisperse but polydisperse. Therefore, micelles are prescribed by the following mass balances:

$$\sum_r N_r = N \quad \text{or} \quad N_r = Y_r N \tag{5.41}$$

$$\sum_r N_i^r N_r = N_i^m \quad \text{or} \quad \bar{N}_i = \sum_r Y_r N_i^r \tag{5.41'}$$

where N_r is the number of the rth micelles (small systems) whose composition is characterized by N_i^r molecules of component i, and Y_r is the fraction of rth micelles against total micelles N. An important point here is that the rth micelles have a definite composition specified by N_i^r. Introducing the above mass balances into (5.17), we obtain

$$dG_t = -S_t\,dT + V_t\,dP + \sum_i \mu_i^s\,dN_i^s + \sum_r \mu_r\,dN_r \tag{5.42}$$

and

$$\mu_r = \sum_i N_i^r\mu_i^m + \xi = (\partial G_t/\partial N_r)_{T,P,N_i^s,N_s} \tag{5.43}$$

For a closed ensemble, $N_i^t(= N_i^s + N_i^m)$ is constant. Thus

$$dG_t = -S_t\, dT + V_t\, dP + \sum_r \left(\mu_r - \sum_i N_i^r \mu_i^s \right) dN_r \qquad (5.44)$$

Therefore, the equilibrium condition at constant T and P becomes

$$\mu_r = \sum_i N_i^r \mu_i^s \qquad (5.45)$$

because the variables N_r are independent. In general, for surfactant solutions in which both micelles and surfactant monomers are dilute, the activities of these components can be replaced by their concentrations with reasonable accuracy:

$$\mu_r = G_r(T, P, N_i^r) + kT \ln C_r \qquad (5.46)$$

$$\mu_i^s = \mu_i^\ominus(T, P) + kT \ln C_i \qquad (5.47)$$

Introducing the above two equations into (5.45), we have

$$K_m = C_r \Big/ \prod_i C_i^{N_i^r} = \exp\left\{ \left(\sum_i N_i^r \mu_i^\ominus - G_r \right) \Big/ kT \right\} \qquad (5.48)$$

This equality indicates that the rth micelle formation can be expressed by the mass-action model. When the ensemble is in equilibrium, the following equality results from (5.45) and (5.46):

$$G_r(T, P, N_i^r) + kT \ln C_r = \sum_i N_i^r \mu_i \qquad (5.49)$$

because $\mu_i^m = \mu_i^s = \mu_i$. In addition, from (5.41) we have

$$Y_r = N_r/N = C_r/C_m \qquad (5.50)$$

where C_m is the total micelle concentration ($C_m = \sum_r C_r$). From (5.49) and (5.50) there results

$$-(H_r/T^2)\, dT + (V_r/T)\, dP - \sum_i N_i^r\, d(\mu_i/T)$$

$$+k\, d \ln Y_r + k\, d \ln C_m = 0 \qquad (5.51)$$

Multiplying (5.51) by Y_r and summarizing over all r, we have

$$-(\bar{H}/T^2)\,dT + (\bar{V}/T)\,dP - \sum_i \bar{N}_i\,d(\mu_i/T) + k\,d\ln C_m = 0 \quad (5.52)$$

On the other hand, if we multiply (5.51) by $Y_r N_k^r$ and summarize over all r, we have

$$d\bar{N}_k = \{(\overline{HN}_k - \bar{H}\bar{N}_k)/kT^2\}\,dT - \{(\overline{VN}_k - \bar{V}\bar{N}_k)/kT\}\,dP$$
$$+(1/k)\sum_i (\overline{N_k N}_i - \bar{N}_k\bar{N}_i)\,d(\mu_i/T) \quad (5.53)$$

At constant T and P, the following equations result, respectively, from (5.52) and (5.53):

$$k\,d\ln C_m = \sum_i \bar{N}_i\,d(\mu_i/T) \quad (5.52')$$

$$d\bar{N}_k = (1/k)\sum_i (\overline{N_k N}_i - \bar{N}_k\bar{N}_i)\,d(\mu_i/T) \quad (5.53')$$

Equation (5.53') also leads to the following equation:

$$d\bar{N} = \sum_k d\bar{N}_k = (1/kT)\sum_i (\overline{NN}_i - \bar{N}\bar{N}_i)\,d\mu_i \quad (5.54)$$

where $N = \sum_k N_k$. Because $x_k = \bar{N}_k/\bar{N}$, the differential of x_k becomes

$$dx_k = (1/\bar{N}kT)\sum_i (\overline{N_i N}_k - \overline{N_i N}x_k)\,d\mu_i \quad (5.55)$$

For two solute components, Eqs. (5.56) and (5.57) are derived respectively from (5.52') and (5.55):

$$kT\,d\ln C_m = \bar{N}_1\,d\mu_1 + \bar{N}_2\,d\mu_2 \quad (5.56)$$

$$\bar{N}kT\,dx_2 = (\overline{N_1 N}_2 - \overline{N_1 N}x_2)\,d\mu_1 + (\overline{N_2^2} - \overline{N_2 N}x_2)\,d\mu_2 \quad (5.57)$$

From (5.56) and (5.57) we have

$$d\mu_1 = \{kT(\overline{N_2^2} - \overline{N_2 N}x_2)\,d\ln C_m - kT\bar{N}_2\bar{N}\,dx_2\}/D \quad (5.58)$$

$$d\mu_2 = -\{kT(\overline{N_1 N}_2 - \overline{N_1 N}x_2)\,d\ln C_m - kT\bar{N}_1\bar{N}\,dx_2\}/D \quad (5.59)$$

where

$$D = \bar{N}(x_1^2\overline{N_2^2} - 2x_1 x_2\overline{N_1 N}_2 + x_2^2\overline{N_1^2}) \quad (5.60)$$

Equations (5.58) to (5.60) give the following relations:

$$(\partial \mu_2 / \partial x_2)_{T,P,C_m} = \bar{N}_1 \bar{N} kT / D \tag{5.61}$$

$$(1/kT)(\partial \mu_1 / \partial \ln C_m)_{T,P,x_2} = (x_1 \overline{N_2^2} - x_2 \overline{N_1 N_2}) / D \tag{5.62}$$

$$(1/kT)(\partial \mu_2 / \partial \ln C_m)_{T,P,x_1} = (x_2 \overline{N_1^2} - x_1 \overline{N_1 N_2}) / D \tag{5.63}$$

Now let us consider the relationships between the total concentration of the ith component (C_i^t) and its monomer concentration (C_i). The variable C_i^t is expressed as

$$C_i^t = C_i + \bar{N}_i C_m \tag{5.64}$$

The differential of C_k^t becomes

$$dC_k^t = dC_k + C_m \, d\bar{N}_k + \bar{N}_k \, dC_m \tag{5.65}$$

Substituting dC_m from (5.52) and $d\bar{N}_k$ from (5.53) into (5.65) leads to the following equation:

$$dC_k^t = (C_m/kT^2)\left(\overline{HN}_k - \sum_i \overline{N_iN}_k h_i^\ominus \right) dT - (C_m/kT)$$

$$\times \left(\overline{VN}_k - \sum_i \overline{N_iN}_k v_i^\ominus \right) dP$$

$$+ C_m \sum_{i \neq k} \overline{N_iN}_k \, d\ln C_i + (C_k + C_m \overline{N_k^2}) \, d\ln C_k \tag{5.66}$$

or, at constant T and P

$$d(\bar{N}_k C_m) = C_m \sum_i \overline{N_iN}_k \, d\ln C_i \tag{5.66'}$$

where the ideal expression $d(\mu_i/T) = k \, d\ln C_i$ at constant T and P is employed. From (5.66) there results

$$(\partial C_k^t / \partial \ln C_k)_{T,P,C_i^t} = C_k + C_m \overline{N_k^2} \tag{5.67}$$

Equations (5.66) and (5.67) are useful for determining the relation between total surfactant concentration and monomer concentration. Micelle aggregation number, which is very useful in dealing with surfactant solutions, is also obtained from these equations.

The following examples are derived from the above discussions. For micelles with one component (component 1), we have the following three equations respectively from (5.52′), (5.53′), and (5.66) at constant T and P:

$$\bar{N}_1 \, d \ln C_1 = d \ln C_m \tag{5.68}$$

$$d\bar{N}_1 = (\overline{N_1^2} - \bar{N}_1^2) \, d \ln C_1 \tag{5.69}$$

$$dC_1^t = (C_1 + C_m \overline{N_1^2}) \, d \ln C_1 \tag{5.70}$$

From the above three equations we have

$$(\partial \ln C_1 / \partial \ln C_m)_{T,P} = 1/\bar{N}_1 \tag{5.71}$$

$$(\partial \bar{N}_1 / \partial \ln C_m)_{T,P} = (\overline{N_1^2} - \bar{N}_1^2)/\bar{N}_1 \tag{5.72}$$

$$(\partial \bar{N}_1 / \partial C_1^t)_{T,P} = (\overline{N_1^2} - \bar{N}_1^2)/(C_1 + C_m \overline{N_1^2}) \tag{5.73}$$

For micelles with two components (1 and 2), we obtain similar equations:

$$\bar{N}_1 \, d \ln C_1 + \bar{N}_2 \, d \ln C_2 = d \ln C_m \tag{5.74}$$

$$dC_1^t = (C_1 + C_m \overline{N_1^2}) \, d \ln C_1 + C_m \overline{N_1 N_2} \, d \ln C_2 \tag{5.75}$$

$$dC_2^t = C_m \overline{N_1 N_2} \, d \ln C_1 + (C_2 + C_m \overline{N_2^2}) \, d \ln C_2 \tag{5.76}$$

Now, from the above three equalities relating the micellar composition, we can derive the relationships among C_1, C_2, C_m, C_1^t, and C_2^t.

5.5. Thermodynamic Functions of Micelle Formation

For a closed ensemble at equilibrium, the following equation can be derived from (5.45) and (5.46) in the same way as (5.39′):

$$-d \, (kT \ln C_r) = -S_r \, dT + V_r \, dP - \sum_i N_i^r \, d\mu_i \tag{5.77}$$

On the other hand, the differential of $d\mu_i$ becomes, from (5.47),

$$d\mu_i = -S_i^\ominus \, dT + V_i^\ominus \, dP + d(kT \ln C_i) \tag{5.78}$$

Substituting (5.78) into (5.77), we have

$$-d(kT \ln C_r) = -\Delta S_r \, dT + \Delta V_r \, dP - \sum_i N_i^r \, d(kT \ln C_i) \qquad (5.79)$$

where

$$\Delta S_r = S_r - \sum_i N_i^r S_i^\ominus \quad \text{and} \quad \Delta V_r = V_r - \sum_i N_i^r V_i^\ominus \qquad (5.80)$$

Thus,

$$\Delta S_r = -\sum_i N_i^r (\partial kT \ln C_i / \partial T)_P + (\partial kT \ln C_r / \partial T)_P \qquad (5.81)$$

and

$$\Delta V_r = \sum_i N_i^r kT(\partial \ln C_i / \partial P)_T - kT(\partial \ln C_r / \partial P)_T \qquad (5.82)$$

When $\sum_i N_i^r$ is very large, say more than 50, the second terms of (5.81) and (5.82) can be neglected compared with the first terms, resulting in

$$\Delta S_r = -\sum_i N_i^r (\partial kT \ln C_i / \partial T)_P \qquad (5.81')$$

$$\Delta V_r = \sum_i N_i^r kT(\partial \ln C_i / \partial P)_T \qquad (5.82')$$

These equations are equivalent to those for the phase-separation model of micelle formation.

In the case of an ionic surfactant, the surfactant ions (1) and counterions (2) are regarded as different components. Then we have from (5.79)

$$-d(kT \ln C_m) = -\Delta S \, dT + \Delta V \, dP - \bar{N}_1 \, d(kT \ln C_1)$$
$$- \bar{N}_2 \, d(kT \ln C_2) \qquad (5.83)$$

At constant temperature, (5.83) is rewritten as

$$\Delta V / \bar{N}_1 = kT(\partial \ln C_1 / \partial P)_T + (\bar{N}_2 kT / \bar{N}_1)(\partial \ln C_2 / \partial P)_T$$
$$- (kT / \bar{N}_1)(\partial \ln C_m / \partial P)_T \qquad (5.84)$$

When $C_1 = C_2 = \text{CMC}$ and the association degree of counterions ($\beta = \bar{N}_2 / \bar{N}_1$) is introduced, (5.84) becomes

$$\Delta V / \bar{N}_1 = (1 + \beta)kT(\partial \ln \text{CMC} / \partial P)_T - (kT / \bar{N}_1)(\partial \ln C_m / \partial P)_T \qquad (5.84')$$

Likewise, there results

$$\Delta S/\bar{N}_1 = -(1+\beta)(\partial kT \ln CMC/\partial T)_P$$
$$+(1/\bar{N}_1)(\partial kT \ln C_m/\partial T)_P \qquad (5.85)$$

5.6. Micellar Parameters Based on Turbidity Data

The interpretation of turbidities in the absence of angular dissymmetry was developed by Stockmayer from the fluctuation of the refractive index in multicomponent systems.[7] The turbidity τ of a solution of C component at constant T and P is given by the equation

$$\tau = HkT \sum_i^c \sum_k^c \rho_i\rho_k(\partial C_i^t/\partial\mu_k)_{\mu_j} \qquad (5.86)$$

where

$$H = 32\pi^3\eta^2/3\lambda^4 N^\circ \qquad (5.87)$$

and η is the refractive index, N° is the number of solvent molecules per unit volume, λ is the wavelength of incident light, $\rho_i = (\partial\eta/\partial C_i^t)_{T,P,C_j}$ and the concentration is the mole ratio of i to solvent. All derivatives in this section are at constant T and P, and these subscripts are therefore omitted. Introducing $(\partial C_i^t/\partial\mu_k)_{\mu_j}$ from (5.66) into (5.86) yields

$$\tau = H\sum_i \rho_i^2 C_i + H\sum_i\sum_k \rho_i\rho_k C_m\overline{N_iN_k} \qquad (5.88)$$

For a one-component micellar solution, we have the following equation from (5.66) and (5.88):

$$\tau = H\rho_1^2(C_1 + \overline{N_1^2}C_m) = H\rho_1^2(\partial C_1^t/\partial \ln C_1)_{T,P} \qquad (5.89)$$

Equation (5.89) leads to

$$\ln C_1 = \ln C_1^0 + \int_{C_1^{t,0}}^{C_1^t} (H\rho_1^2/\tau)\, dC_1^t \qquad (5.90)$$

On the other hand, (5.91) is derived from (5.52′) at constant T and P for one component

$$\mathrm{d} \ln C_m = \bar{N}_1 \, \mathrm{d} \ln C_1 \tag{5.91}$$

The following equation is derived from (5.89) and (5.91):

$$\mathrm{d}(C_1 + C_m) = (H\rho_1^2/\tau)C_1^t \, \mathrm{d}C_1^t \tag{5.92}$$

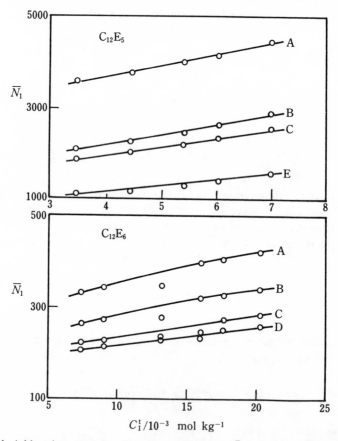

$C_1^t/10^{-3}$ mol kg^{-1}

Figure 5.1. Arithmetic mean aggregation number of micelles ($\bar{N}_1$) of pentaoxyethylene dodecyl ether ($C_{12}E_5$) and hexaoxyethylene dodecyl ether ($C_{12}E_6$) at 298.15 K as a function of total surfactant concentration (C_1^t) under various pressures.[5] A, 0.1 MPa; B, 20 MPa, C, 40 MPa; D, 80 MPa; E, 100 MPa. (Reproduced with permission of the American Chemical Society.)

where the mass balance of (5.64) is employed. Then, we have

$$C_1 + C_m = C_1^0 + C_m^0 + \int_{C_1^{t,0}}^{C_1^t} (H\rho_1^2/\tau)C_1^t \, dC_1^t \tag{5.93}$$

Now, from the turbidity change with total concentration, C_1, C_m, $\bar{N}_1$, and $\overline{N_1^2}$ are evaluated from Eqs. (5.88) to (5.93). In (5.93), it can be assumed that $C_1^0 = C_1^{t,0} = CMC$ and $C_m^0 = 0$ in applications of turbidity experiments.

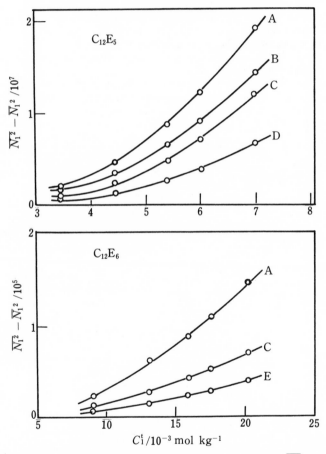

Figure 5.2. Variance of aggregation number distribution of micelles $(\overline{N_1^2} - \bar{N}_1^2)$ of pentaoxyethylene dodecyl ether ($C_{12}E_5$) and hexaoxyethylene dodecyl sulfate ($C_{12}E_6$) at 298.15 K as a function of total surfactant concentration (C_1^t) under various pressures.[5] A, 0.1 MPa; B, 20 MPa; C, 40 MPa; D, 100 MPa; E. 160 MPa. (Reproduced with permission of the American Chemical Society.)

The experimental results obtained for two nonionic surfactants—penta- and hexaoxyethylene dodecyl ethers ($C_{12}E_5$ and $C_{12}E_6$)—are illustrated in Figs. 5.1 and 5.2 for the mean micelle aggregation number and its distribution, respectively. The effect of pressure on these quantities is also illustrated.

References

1. T. L. Hill, *Thermodynamics of Small Systems*, Vols. 1 and 2, Benjamin, New York (1963, 1964).
2. D. G. Hall and B. A. Pethica, in: *Nonionic Surfactants* (M. Schick, ed.), pp. 516–557, Dekker, New York (1967).
3. D. G. Hall, *Trans. Faraday Soc.* 66, 1351, 1359 (1970).
4. D. G. Hall, in: *Aggregation Processes in Solution* (E. Wyn-Jones and J. Gormally, eds.), pp. 7–69, Elsevier, Amsterdam (1983).
5. M. Okawauchi, M. Shinozaki, Y. Ikawa, and M. Tanaka, *J. Phys. Chem.* 91, 109 (1987).
6. I. Prigogine and R. Defay, *Chemical Thermodynamics*, Chapter 1, Longmans, New York (1967).
7. W. H. Stockmayer, *J. Phys. Chem.* 18, 58 (1950).

6

Micelle Temperature Range (MTR or Krafft Point)

6.1. Introduction

Detergent action, colloid formation, and surface activity are different manifestations of the same characteristics of surfactant solutions. Detergency is the most important and conventional function of soaps, and is closely connected with their solubility. Because typical surfactant molecules (including soaps) have both a hydrophilic group and a bulk hydrophobic group, their aqueous solubility is not expected to be high. That is true below a certain narrow temperature range called the *Krafft point*. Above this range, the solubility of the surfactant increases very steeply because surfactant aggregation is taking place.

In the previous chapters, the dissolution and micellization of surfactants in aqueous solutions were discussed from the standpoint of the degrees of freedom as given by the phase rule. The mass-action model for micelle formation was found to be better for explaining the phenomena of surfactant solutions than the phase-separation model. Two models have similarly been used to explain the Krafft point, one postulating a phase transition at the Krafft point and the other a solubility increase up to the CMC at the Krafft point. The most recent version of the first approach is a melting-point model for a hydrated surfactant solid.[1,2] The most direct approach to the second model of the Krafft point rests entirely on measurements of the solubility and CMC of surfactants with temperature. From these mesurements the concept of the Krafft point can be made clear. This chapter first reviews the concepts used to relate the dissolution of surfactants to their micellization, and then shows that the concept of a micelle temperature range (MTR) can be used to elucidate various phenomena concerning dissolution and micellization of surfactants not only in relation to the phase rule but also experimentally.

6.2. Krafft Point and Related Technical Terms

In 1895, Krafft and Wiglow published a paper on soap solutions entitled, "On the behavior of fatty acid alkalies and soaps in the presence of water."[3] They coined the term *Ausscheidungstemperatur* for the temperature at which a new phase separates from a soap solution on cooling. They found that this temperature was lower than the melting point of the corresponding fatty acids, and that, for the six fatty acids examined, the difference between the separation temperature and the melting point increased with decreasing alkyl chain length of the fatty acids. In 1926, McBain and Elford published a phase diagram for a potassium oleate-water system that showed the minimum temperatures at which heterogeneous systems turned to homogeneous isotropic liquids during heating.[4] These homogeneous solutions separated again at the same temperature on cooling. Lawrence in 1935 introduced the term *Krafft point*, which he interpreted as marking a phase transition[5]: "The Krafft point is then due to loosening of the attractive forces between hydrocarbon chains throughout the micelle. This is a phase change of their adhesion; from solid to liquid. The degree of dispersion is thereby greatly increased." In the same year, Murray and Hartley used the mass action equation to explain a rapid solubility increase over a narrow temperature range.[6]

A new concept of the Krafft point soon emerged. In 1951, Eggenberger and Harwood performed conductometric studies on the solubility and micelle formation of dodecylammonium chloride, and found a sharp break in the solubility versus temperature curve that could be attributed to the "Krafft effect," i.e., to solubilization of the undissociated molecule by the micelle. Significantly, the CMC versus temperature curve intersected the precipitation curve at this point.[7] The sharpness of this break in the precipitation curve was interpreted as indicating that micelles do not form appreciably below the CMC. This view of the Krafft point has become established. Phillips defined the Krafft point as the temperature at which the CMC is equal to the saturation solubility.[8] Alexander and Johnson interpreted the Krafft point phenomenon as the point at which the transfer of soap molecules "is effective from the solid phase to the micelle, since the concentration of single molecules only increases quite slowly once micelles are present."[9] This concept is totally correct. Shinoda and Hutchinson, on the other hand, suggested that the Krafft point represents a freezing of the micelle or a melting of hydrated solids of surfactants: "the Krafft point now can be interpreted as a point at which solid hydrated agent and micelles are in equilibrium with monomers: in terms of phase, with two components the equilibrium hydrated solid $\rightleftharpoons$ monomers $\rightleftharpoons$ micelles is univariant, so that at a given pressure the point is fixed."[1] Unfortunately, this concept is incorrect, as will be made clear in the following sections.

Another term, the *critical solution temperature* (CST), was introduced to designate the temperature beyond which the solubility of nonionic surfactants in organic solvents increases markedly, as marked by an inflection in the solubility curve.[10] Mazer and Benedek used the *critical micellar temperature* (CMT) to refer to the phase boundary between a hydrated solid phase and a micellar phase.[11] The CMT value was taken as the midpoint of the temperature range over which the hydrated solid phase clarified on slow warming with vigorous shaking.

As can be seen, concepts of the Krafft point fall ito two different categories: those involving a phase transition of solid surfactant and those involving a solubility increase up to the CMC. The two views are incompatible: according to the former, the Krafft point is a definite temperature (a point), whereas according to the latter, it is a narrow temperature range. This conflict is addressed in the following sections.

6.3. The Physicochemical Meaning of the MTR

The Krafft point has been defined as the temperature at which the solubility versus temperature curve intersects the CMC versus temperature curve. Let us think about this definition by using Fig. 6.1.[12] If the micelle is regarded as a phase, three phases (intermicellar bulk phase, surfactant solid phase, and micellar phase) coexist at the Krafft point. The number of components is two (water and surfactant), so the Gibbs phase rule ($f = C - P + 2$) gives only one degree of freedom (f, C, and P are the number of degrees of freedom, component, and phase, respectively). Thus, specifying the pressure automatically determined the Krafft point T_k. From Fig. 6.1, this conclusion may seem to apply well and to be consistent with the phase rule. However, the same three phases still coexist at every point of the solubility curve above the Krafft point—at point P for example. In other words, at a constant pressure of 1 atm, innumerable temperatures on the curve are in equilibrium with the three phases, which is contrary to the conclusion derived from the phase-separation model. Thus, the phase-separation model cannot be applied to typical micelles with aggregation numbers less than a few hundred. This model does apply if the aggregation number is infinite: in that case, a phase separation would take place and the solubility curve would approach the path $P_0 \rightarrow P_K \rightarrow P'$ with increasing

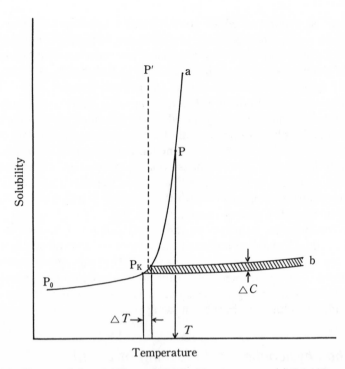

Figure 6.1. Changes of the solubility and CMC with temperature: (a) Solubility curve, (b) CMC curve; (ΔC, narrow concentration range of CMC; ΔT, micelle temperature range). (Reproduced with permission of Academic Press.)

temperature, with the monomer concentration remaining constant at S_{T_k}. Thus, one temperature T_k would be specified at 1 atm pressure.

When the mass-action model is applied to micellization, on the other hand, the following association equilibrium between surfactant monomers (S) and micelles (M) can be considered:

$$K_n = [M]/[S]^n \qquad (6.1)$$

where K_n is the equilibrium constant of micellization and n is the aggregation number of micelles. In this case, there are two degrees of freedom ($f = C - P + 2 - r$) because there are two phases (surfactant solution phase and surfactant solid phase), three components (water, monomeric surfactant, and micelle), and one equilibrium equation, where r is the number of equilibrium equations. Then, any two intensive thermodynamic varibles— temperature and pressure, for example—can specify the system of the solution. This conclusion agrees completely with experimental findings that the solubility is determined only by temperature at atmospheric pressure. The simplified expression (6.1) may not be adequate to express the general micelle formation of ionic surfactants, because counterions and the size distribution of micelles are not taken into account, but it is satisfactory for discussing the degrees of freedom (see Chapter 4).

Another approach to the Krafft point is the melting-point model of a hydrated solid surfactant.[1,13-15] In this model, the solubility curve represents the hydrated surfactant solid below the Krafft point and the melted surfactant phase above it. If the micelle is regarded as a phase, the system must be invariant ($f = 0$) at the Krafft point because four phases (intermicellar bulk phase, hydrated surfactant solid phase, melted surfactant phase, and micellar phase) coexist for two components. However, experimental evidence shows that the CMC changes with pressure.[16,17]

Even if the micelle is regarded as a chemical species, the melting-point model is not correct. In this case, the system is monovariant ($f = 1$), and the Krafft point is determined automatically at 1 atm pressure. That seems reasonable for a single surfactant solution, but when the model is applied to a mixed surfactant solution, it is found to be incorrect. Figure 6.2 shows the phase diagram of a water/sodium dodecyl sulfate (SDS)/manganese (II) dodecyl sulfate [Mn (DS)$_2$] system, with temperature as the ordinate.[18] The CMC of the surfactant mixture gives a curved surface between the

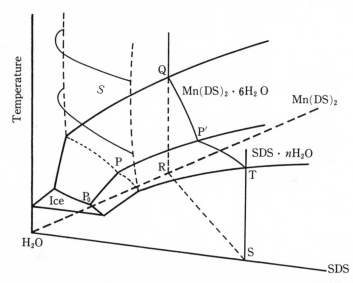

Figure 6.2. Phase diagram of water/sodium dodecyl sulfate/manganese(II) dodecyl sulfate system. *S*, mixed CMC surface. (Reproduced with permission of Academic Press.)

CMC–temperature curves of each component. By analogy with the two-component phase diagram, we can conclude from this three-component phase diagram that the rational Krafft point for a binary surfactant mixture is determined by the intersection of the mixed CMC surface with both humped surfaces of the surfactant solids.

Consider point P in Fig. 6.2. If the melting-point model is applied, five phases coexist at point P [micellar solution, SDS solid, melted SDS, $Mn(DS)_2$ solid, melted $Mn(DS)_2$], and the number of degrees of freedom is zero. If this were true, phase diagrams like Fig. 6.2 could be drawn only at 1 atm pressure. In fact, the solubility and CMC of surfactants have been measured from one to several thousand atmospheres.[17,19]

The surface QRSTP′ on Fig. 6.2, formed by the intersection between the humped surfaces of the two surfactant solids and a plane of constant total concentration far above the CMC, looks very similar to a phase diagram of two components with a eutectic. This similarity of shape seems to have led to the model of melting of a hydrated surfactant solid, and also to a thermodynamic analysis of the Krafft point of a binary surfactant mixture based on an analogy with the freezing point depression of a binary mixture.[20] This model gives an absurd conclusion, however, because it neglects the presence of the third component, water, which is essential to any discussion of micelle formation. Moreover, if a micelle is regarded as a phase, there are −1 degrees of freedom—also absurd.

If a micelle is regarded as a chemical species and no melting of the surfactant solids is assumed, two degrees of freedom still remain along the line of $P_0 \rightarrow P \rightarrow P'$ (because there are four components, three phases, and one equilibrium equation). Temperature therefore specifies the binary surfactant system at a definite pressure as long as two surfactant solid phases coexist in the system, which is consistent with much experimental evidence.

Bivalent metal dodecyl sulfates or sulfonates are typical hydrated surfactant solids, for example, $Cu(DS)_2 \cdot 4H_2O$ and $Cu(DSo)_2 \cdot 2H_2O$. For both of these compounds, the Krafft point and the phase transition temperature are not the same (Krafft points 19.0^{18} and $53.5°C^{21}$, respectively; phase transition temperatures 44 and 66°C, respectively[12]), indicating that the melting-point model is wrong. There are cases, however, in which the phase transition temperature of the hydrated surfactant solid happens to be very near the Krafft point of the surfactant.[22,23]

It can be concluded that the Krafft point is the temperature at which the solubility of surfactants as monomers becomes high enough for the monomers to commence aggregation or micellization. Recall from Chapter 4 that the CMC depends on the method used for its determination, and that the CMC value should therefore be defined as a narrow temperature range[24] even though the solubility is definitely determined by temperature alone under conditions of constant pressure. This fact results from the polydispersity of micelles, as discussed in Chapter 4. The *micelle temperature range* (MTR) may now be defined as the temperature range delimited by the intersection of the solubility versus temperature curve with the CMC range versus temperature curve even though each CMC determination method yields a single, definite MTR temperature. In other words, the MTR is determined by the balance between CMC and solubility and by the dependence of both on temperature.[21,25]

Now that the definition of the Krafft point has been made clear, it is appropriate to make a few remarks about how to shift the MTR to a lower temperature, as this is of considerable practical importance for ionic surfactants. From the above definition, it should be clear that there are two ways in principle to decrease the MTR: by increasing the monomer solubility of the surfactant while leaving the CMC unchanged (process *a* in Fig. 6.3); or by decreasing the CMC while leaving the monomer solubility unchanged (process *b* in Fig. 6.3).[12] The first method is closely related to the crystal state of the surfactant: the less energetically stable the solid surfactant, the higher its solubility. The stability of the solid can be lowered by such

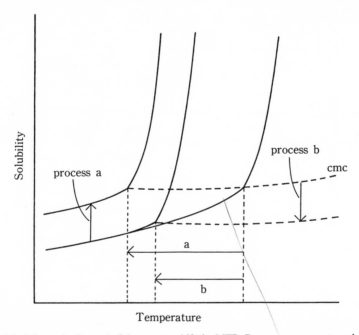

Figure 6.3. Schematic diagram of the ways to shift the MTR. Process *a* represents an increase of monomer solubility; process *b* represents a decrease of the CMC value. (Reproduced with permission of Academic Press.)

manipulations as dispersing the electrical charge of the counterion, increasing the counterion volume, increasing the water of crystallization, and introducing a branched chain into the hydrophobic moiety. Alternatively, a decrease in MTR linked to a decrease in CMC could be brought about by increasing association of counterion to micelle, increasing the hydrophobicity of counterion and surfactant ion, and so on.

It is now easy to understand the changes in MTR obtained by manipulating the aqueous solubility and CMC of tetradecane-1-sulfonates $[C_n BP(C_{14})_2]$ with the divalent cationic counterion 1, $1' - (1,$ ω-alkanediyl)bispyridinium

$$\langle\!\langle\bigcirc\rangle N^+ - (CH_2)_n - {}^+N\langle\bigcirc\rangle\!\rangle$$

$$n = 2, 4, 6, 8, 10, 12, 14$$

with a separated electric charge (Figs. 6.4 and 6.5; Table 6.1).[26,27] The

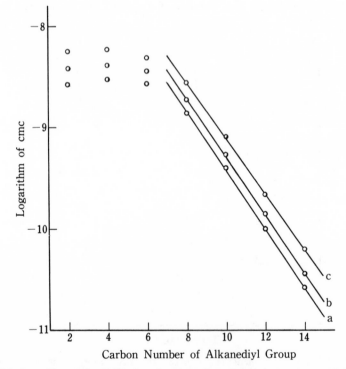

Figure 6.4. Logarithm of the CMC plotted against alkanediyl chain length at 35°C (a), 45°C (b), and 55°C (c). (Reproduced with permission of Academic Press.)

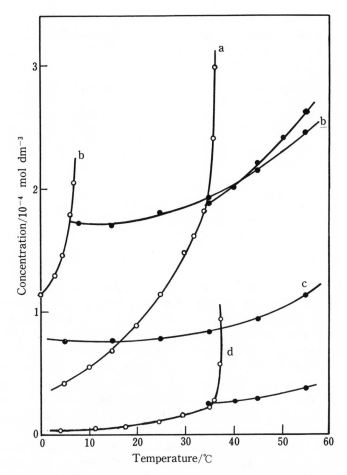

Figure 6.5. Changes in solubility (○) and CMC (●) with temperature: (a) $C_2BP(C_{14})_2$, (b) $C_6BP(C_{14})_2$, (c) $C_{10}BP(C_{14})_2$, and (d) $C_{14}BP(C_{14})_2$. (Reproduced with permission of Academic Press.)

decrease in the MTR with increasing alkanediyl carbon number from $C_2BP(C_{14})_2$ to $C_4BP(C_{14})_2$ to $C_6BP(C_{14})_2$ evidently results from process *a* above, because the solubilities increase while the CMC values remain almost the same. The further decreases in MTR for $C_8BP(C_{14})_2$ and $C_{10}BP(C_{14})_2$ are due to both solubility increase and CMC decrease (processes *a* and *b* combined). The increase of MTR through the minimum for $C_{12}BP(C_{14})_2$ and $C_{14}BP(C_{14})_2$ clearly results from the pronounced stability of the solid surfactants (i.e., their low solubility). The decrease in CMC for these two surfactants contributes to the decrease in MTR. In this case, the effect of

Table 6.1. Micelle Temperature Range (MTR) and
Heat of Dissolution (Δh^0) of 1, 1'-[1, ω − alkanediyl]-
bispyridinium tetradecane-1-sulfonate[a]

Surfactant	MTR (°C)	Δh^0 (kJ · mol^{-1})
$C_2BP(C_{14})_2$	34.6	106
$C_4BP(C_{14})_2 \cdot 2H_2O$	11.4	98
$C_6BP(C_{14})_2$	6.0	80
$C_8BP(C_{14})_2$	<0	—
$C_{10}BP(C_{14})_2$	<0	—
$C_{12}BP(C_{14})_2$	2.1–2.3	—
$C_{14}BP(C_{14})_2$	35.7	133

[a]Reproduced with permission of Academic Press.[26]

decreased solubility on the MTR increase is much greater than the effect of a decrease in CMC on the MTR decrease, judging from the fact that MTR depends absolutely on the balance between the monomeric solubility and the CMC of surfactants.

The following general conclusions can be drawn from the above CMC, solubility, and MTR values and from Δh^0. First, divalent counterions with separated electric charge can move readily over the charged micelle surface as long as the charge separation is small. Divalent counterions with concentrated (e.g., Cu^{2+}) or diffuse (e.g., methylviologen, MV^{2+}) charge show the same behavior.[21,28] Counterions with divalent charges separated by more than six methylene groups, on the other hand, become anchored to the micellar surface, leading to a decrease in the CMC. Second, the surfactant solids with a divalent counterion initially become less stable (more soluble) as the charge separation is increased, but beyond a certain charge separation become more stable (less soluble).

Pressure is another thermodynamic parameter that affects the MTR, because both the monomer solubility and the CMC are pressure dependent. Nevertheless, only a few reports have appeared concerning the effects of pressure.[29-31] Changes in CMC with pressure exhibit a shallow maximum, as was seen in Chapter 4. The CMC is less pressure dependent than the solubility. Solubility decreases rapidly with increasing pressure because dissolution involves a positive volume change. Therefore, the difference between CMC and solubility at a given temperature is accentuated by pressure, leading to a higher MTR (Fig. 6.6). At high pressure, therefore, the solubility increases up to the CMC.

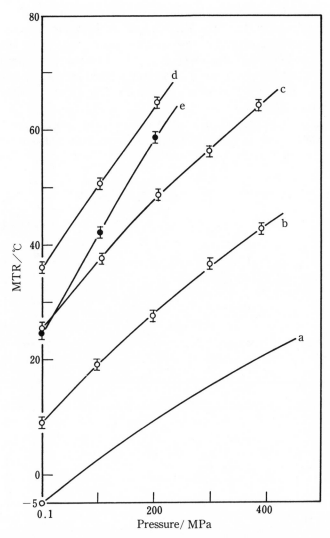

Figure 6.6. Changes in the MTR with pressure for typical ionic surfactants: (a) sodium decyl sulfate, (b) sodium dodecyl sulfate, (c) sodium tetradecyl sulfate, (d) sodium hexadecyl sulfate, (e) hexadecyltrimethylammonium bromide. (Reproduced with permission of the American Chemical Society.)

6.4. MTR Change of Homologous Surfactants

The definition of the MTR adopted above can be used to explain several experimental facts. The Krafft points of homologous ionic surfactants have been reported to increase with increasing alkyl chain length.[32–36] The important feature of this increase is that MTR values do not increase linearly with alkyl chain carbon number, but instead increase gradually and then plateau (Fig. 6.7). The curve shows some irregularities related to even versus odd alkyl chain carbon numbers,[32,33] a difference that affects the solid structure of the substances and therefore such physicochemical properties as solubility, melting point, and the long spacing of the crystal lattice.[37,38] As shown in Fig. 6.8, the logarithm of both solubility and CMC decreases

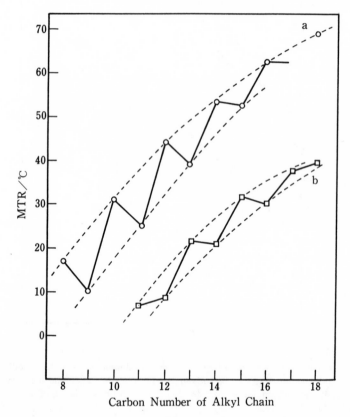

Figure 6.7. Changes in the MTR with the carbon number of the surfactant alkyl chain: (a) sodium soaps of fatty acids, (b) sodium alkyl sulfates. (Reproduced with permission of the Chemical Society of Japan.)

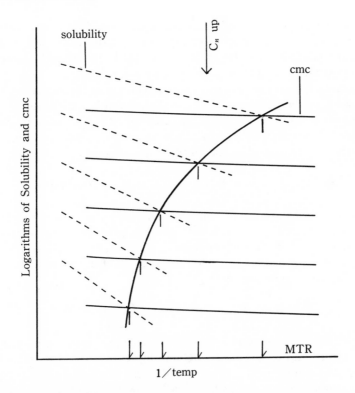

Figure 6.8. Schematic illustration of the changes in the MTR with increasing carbon number of surfactant ion. (Reproduced with permission of the Chemical Society of Japan.)

linearly with increasing carbon number in an odd or even series,[21,35] where the logarithm of solubility is assumed to be less than that of the CMC at lower temperature. In addition, the heat of dissolution below the MTR becomes more positive with increasing carbon number of the surfactant ion,[21] i.e., the rate of solubility increase with temperature is more rapid for surfactant ions with longer alkyl chains. Conversely, the change in CMC with temperature is not strongly affected by alkyl chain length. Using the definition of MTR based on the intersection of the solubility and CMC curves, the leveling of MTR in Fig. 6.8 is easily understood.

6.5. MTR Change with Additives

Adding salts with counterions to an ionic surfactant solution decreases the CMC and increases the micellar aggregation number. Excess counterions also decrease the solubility of surfactants below the MTR so as to maintain a constant solubility product of counterions and surfactant ion, and thus change the MTR. In fact, the MTR of SDS can be increased by increasing the concentration of sodium chloride in solution.[11,39]

Because the aim of this section is to show that the definition of the MTR given above can be used to explain observed deviations in the MTR, the following calculation for estimating the CMT of a surfactant is adequate.[11] From Eq. (4.38), the relation between the CMC and counterion concentration for monodisperse micelle formation is

$$\ln \text{CMC} = -(m/n) \ln[\text{G}] + (1/n) \ln K_n + (1/n) \ln[\text{M}_n] \qquad (6.2)$$

because the concentration of monomeric surfactant ions at the CMC can be replaced by the total surfactant concentration with more than 99% confidence for $n > 50$ (see Chapter 4). In most cases ($n > 50$), the last term on the right-hand side is less than a few percent of the constant term, and (6.2) is well approximated by

$$\ln \text{CMC} = -(m/n) \ln[\text{G}] + \text{constant} \qquad (6.3)$$

where m/n is the degree of counterion association with the micelle. The linear relationship between $\ln \text{CMC}$ and $\ln[\text{G}]$ has been ascertained in a number of experiments.[40,41] The CMC of SDS is 8.4×10^{-3} mol $\cdot$ dm^{-3}, and m/n is extrapolated to be 0.72 to 0.75[41] at an MTR of 9°C. The value of the constant in (6.3) becomes -8.22 and -8.36 for m/n value of 0.72 and 0.75, respectively. On the other hand, the solubility of SDS decreases with increasing counterion concentration. Keeping the solubility product constant at 9°C,

$$C(C + C_{\text{Na}^+}) = (8.4 \times 10^{-3})^2 \qquad (6.4)$$

where C is the concentration of surfactant or solubility. From (6.3) and (6.4), the difference between CMC and C at 9°C can be evaluated at different C_{Na^+} values. The change in CMC with temperature is negligible compared with the change in solubility. The MTR can then be determined as the temperature at which the difference between the above two concentrations drops to zero as a result solely of the increase in solubility with temperature. Figure 6.9 shows the MTR range obtained in this way from the two m/n values given above, together with the experimental results.[11] The heat of dissolution used for the calculation is 50 kJ $\cdot$ mol^{-1}.[13] The difference of less than a few degrees between MTR and CMT, as demonstrated by Mazer and Benedek,[11] is quite reasonable judging from the definition of CMT.

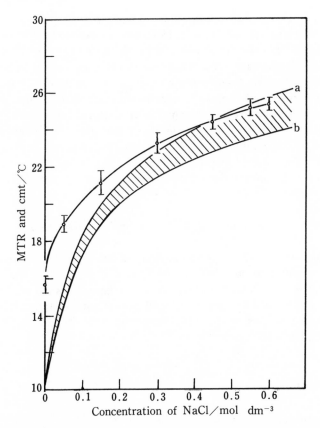

Figure 6.9. MTR increase of sodium dodecyl sulfate with increasing sodium chloride concentration: ♀ CMC change; (a) $m/n = 0.72$, (b) $m/n = 0175$. (Reproduced with permission of the Chemical Society of Japan.)

Inorganic additives do not always elevate the Krafft point. The Krafft points of zwitterionic surfactants are depressed by the addition of salts.[20]

Organic additives also affect the MTR. Shirahama *et al.* examined the effects of a number of water-soluble organic additives on the CMC and solubility of anionic surfactants, and discussed the Krafft points resulting from the balance between the two.[42,43] When the effect on the CMC decrease was greater than the effect on the solubility change, the MTR decreased. This was the case for methanol, ethanol, propranol, acetone, and dioxane. Shinoda *et al.* repoted typical examples of MTR depression owing to CMC decrease for some higher alcohols (C_6 to C_8),[13] although they used a definition of the Krafft point based on the melting point of a hydrated surfactant solid.

Much effort has been put into finding ways to depress the Krafft points of ionic surfactants, for example by introducing oxyethylene[14,44] or oxypropylene groups[45] between the hydrocarbon chain and the ionic head group. In this case, the decrease in the Krafft points is brought about by a decrease in the CMC rather than an increase in surfactant solubility. Another area of interest is the Krafft points of fluorinated surfactants, which are much higher than those of corresponding hydrocarbon surfactants, largely because of the low aqueous solubility of these compounds.[13,46] The effect of the solvent on the solubility and CMC is also of interest. Substitution of formamide for water greatly increases the CMCs of SDS and hexadecyltrimethylammonium bromide, resulting in a higher MTR (55 and 43°C, respectively).[47] The review article by Sowada on the Krafft point is very suggestive.[48]

The above discussion leads to the following conclusions.

1. The term *micelle temperature range* expresses the relation between solubility and micelle formation better than *Krafft point.*
2. The MTR is determined only by the balance between the CMC and solubility and the dependence of both values on temperature.
3. These relations can be explained perfectly by the mass-action model of micelle formation.[25]

References

1. K. Shinoda and E. Hutchinson, *J. Phys. Chem. 66*, 577 (1962).
2. K. Shinoda, *J. Phys. Chem. 85*, 3311 (1981).
3. F. Krafft and H. Wiglow, *Ber. Dtsch. Chem. Ges. 28*, 2566 (1895).
4. J. M. McBain and W. J. Elford, *J. Chem. Soc. 129*, 421 (1926).
5. A. S. C. Lawrence, *Trans. Faraday Soc. 31*, 206 (1935).
6. R. C. Murray and G. S. Hartley, *Trans. Faraday Soc. 31*, 183 (1935).
7. D. N. Eggenberger and H. J. Harwood, *J. Am. Chem. Soc. 73*, 3353 (1951).
8. J. N. Phillips, *Trans. Faraday Soc. 51*, 561 (1955).
9. A. E. Alexander and P. Johnson, in: *Colloid Science*, pp. 683–685, Oxford University Press, London (1949).
10. K. Kon-no, T. Jin-no, and A. Kitahara, *J. Colloid Interface Sci. 49*, 383 (1974).
11. A. M. Mazer and G. B. Benedek, *J. Phys. Chem. 80*, 1075 (1976).
12. Y. Moroi, R. Sugii, and R. Matuura, *J. Colloid Interface Sci. 98*, 184 (1984).
13. H. Nakayama, K. Shinoda, and E. Hutchinson, *J. Phys. Chem. 70*, 3502 (1966).
14. M. Hatao and K. Shinoda, *J. Phys. Chem. 77*, 378 (1973).
15. K. Tsujii, N. Saito, and T. Takeuchi, *J. Phys. Chem. 84*, 2287 (1980).
16. R. F. Tuddenham and A. E. Alexander, *J. Phys. Chem. 66*, 1839 (1962).
17. S. Kaneshina, M. Tanaka, T. Tomida, and R. Matuura, *J. Colloid Interface Sci. 48*, 450 (1974).
18. Y. Moroi, T. Oyama, and R. Matuura, *J. Colloid Interface Sci. 60*, 103 (1977).

19. M. Tanaka, S. Kaneshina, S. Kuramoto, and R. Matuura, *Bull. Chem. Soc. Jpn. 48*, 432 (1975).
20. K. Tsujii and J. Mino, *J. Phys. Chem. 82*, 1610 (1978).
21. Y. Moroi, R. Sugii, C. Akine, and R. Matuura, *J. Colloid Interface Sci. 108*, 180 (1985).
22. M. Kodama and S. Seki, *Prog. Colloid Polym. Sci. 68*, 158 (1983).
23. M. Kodama and S. Seki, *Netsu Sokutei 11*, 104 (1984).
24. W. C. Preston, *J. Phys. Chem. 52*, 84 (1948).
25. Y. Moroi and R. Matuura, *Bull. Chem. Soc. Jpn. 61*, 333 (1988).
26. Y. Moroi, R. Matuura, T. Kuwamura, and S. Inokuma, *J. Colloid Interface Sci. 113*, 225 (1986).
27. R. Matuura, Y. Moroi, and M. Ikeda, in: *Surfactants in Solution* 4 (K. L. Mittel and P. Bothorel, eds.), pp. 289-298, Plenum Press, New York (1986).
28. Y. Moroi, N. Ikeda, and R. Matuura, *J. Colloid Interface Sci. 101*, 285 (1984).
29. M. Tanaka, S. Kaneshina, T. Tomida, K. Noda, and K. Aoki, *J. Colloid Interface Sci. 44*, 525 (1973).
30. N. Nishikido, H. Kobayashi, and M. Tanaka, *J. Phys. Chem. 86*, 3170 (1982).
31. H. W. Offen and W. D. Turley, *J. Colloid Interface Sci. 92*, 575 (1983).
32. H. Lange and M. J. Schwuger, *Kolloid Z. Z. Polym. 223*, 145 (1967).
33. K. Ogino and Y. Ichikawa, *Bull. Chem. Soc. Jpn. 49*, 2682 (1976).
34. K. Shinoda, Y. Minegishi, and H. Arai, *J. Phys. Chem. 80*, 1987 (1976).
35. M. Saito, Y. Moroi, and R. Matuura, *J. Colloid Interface Sci. 88*, 758 (1982).
36. M. Saito, Y. Moroi, and R. Matuura, in: *Surfactants in Solution* 2 (K. L. Mittal and B. Lindman, eds.), pp. 771-788, Plenum Press, New York (1984).
37. A. E. Bailey, in: *Melting and Solidification of Fats*, p. 184, Interscience, New York (1950).
38. A. Muller, *Proc. R. Soc. London. Ser. 114A*, 542 (1927).
39. H. Nakayama and K. Shinoda, *Bull. Chem. Soc. Jpn. 40*, 1797 (1967).
40. M. L. Corrin and W. D. Harkins, *J. Am. Chem. Soc. 69*, 683 (1947).
41. Y. Moroi, N. Nishikido, H. Uehara, and R. Matuura, *J. Colloid Interface Sci. 50*, 254 (1975).
42. K. Shirahama, M. Hayashi, and R. Matuura, *Bull. Chem. Soc. Jpn. 42*, 1206 (1969).
43. K. Shirahama, M. Hayashi, and R. Matuura, *Bull. Chem. Soc. Jpn. 42*, 2123 (1969).
44. M. Hato, M. Tahara, and Y. Suda, *J. Colloid Interface Sci. 72*, 458 (1979).
45. K. Shinoda, M. Maekawa, and Y. Shibata, *J. Phys. Chem. 90*, 1228 (1986).
46. K. Shinoda, M. Hato, and T. Hayashi, *J. Phys. Chem. 76*, 909 (1972).
47. I. Rico and A. Lattes, *J. Phys. Chem. 90*, 5870 (1986).
48. R. Sowada, *Chem. Technol.* (Leipzig), *37* 470 (1985).

7

Stability of Colloidal Particles

7.1. The Debye–Hückel Theory

The Debye–Hückel theory is covered in almost all textbooks on electrolyte solutions,[1,2] but it will be reviewed in this chapter as a tool for modeling the interaction between charged colloidal particles of larger size. Because electrolytes in aqueous solution dissociate into ionic species that interact electrostatically, the concentration dependence of the activity coefficient differs sharply for electrolyte and nonelectrolyte solutions. This section is devoted to estimating the interaction between ionic species and deriving their activity coefficients. Three assumptions are adopted: (1) a solution is a dielectric continuum of constant ε; (2) ions are hard spheres of diameter a; and (3) the concentration is relatively low (at higher concentrations the Debye–Hückel theory is very approximate). Consider a charge density in a volume element dv at a distance r from an arbitrarily selected central ion and assume the mean electrostatic potential to be ψ_r. The Poisson equation is used to relate the charge density in dv to the electrical potential ψ_r:

$$\nabla \psi_r = -4\pi\rho/\varepsilon \tag{7.1}$$

where ∇ is the Laplacian.

   There is no doubt that the mean distribution of positively and negatively charged ions around the central ion is spherically symmetrical, provided no additional forces are acting on the ions. The distribution simply represents the time-averaged effects of the mutual interaction and thermal motion of the ions. Therefore, the Laplacian turns into the following simple form:

$$(1/r^2) \times d(r^2\, d\psi_r/dr)/dr = -4\pi\rho/\varepsilon \tag{7.2}$$

The charge density in the volume element is equal to the excess charge in the volume element, which is equal to the sum of each ion density n_i (the average number of i ions per unit volume) times the charge $z_i e$ on the ion:

$$\rho = \sum_i n_i z_i e \tag{7.3}$$

where e is the electronic charge. For the central ion, the following equation is satisfied from the electroneutrality condition of solution:

$$\int_{r_0}^{\infty} 4\pi r^2 \rho \, dr = -ze \tag{7.4}$$

where ze is the charge of the central ion and r_0 is the distance at closest approach between the central ion and the surrounding ions. On the other hand, the ion density n_i in (7.3) is related to the electrical potential by Boltzmann's theorem

$$n_i = n_i^0 \exp(-z_i e\psi_r / kT) \tag{7.5}$$

where n_i^0 is the average bulk density of i ion at the point where $\psi_r = 0$. This is the first Debye–Hückel approximation. Now, the charge density ρ_r, which is a function of r of the volume element, becomes

$$\rho_r = \sum_i n_i^0 z_i e \exp(-z_i e\psi_r / kT) \tag{7.6}$$

As mentioned above, there is no additional force applied to the system, and therefore the kinetic energy of ions owing to their thermal motion is expected to be much greater than their electrostatic energy, i.e., $z_i e\psi_r / kT \ll 1$ (otherwise, complete dissociation of a strong electrolyte could not occur in a dilute solution). This is the second Debye–Hückel approximation, which is justified when the solution is sufficiently dilute. Thus, the exponential in (7.6) can be expanded in a linear Taylor series, and the Poisson–Boltzmann equation finally becomes

$$(1/r^2) \times d(r^2 \, d\psi_r / dr)/dr = \kappa^2 \psi_r \tag{7.7}$$

and

$$\kappa^2 = (4\pi e^2 / \varepsilon kT) \times \sum_i n_i^0 z_i^2 \tag{7.8}$$

where the electroneutrality condition of solution is employed:

$$\sum_i n_i^0 z_i e = 0 \tag{7.9}$$

The symbol κ is not only a shorthand symbol but also indicates a very important parameter concerning the distribution of ions around the central ion. The value of κ^2 is proportional to ionic strength, and becomes zero as the solution approaches an infinite dilution. The linearized Poisson-Boltzmann equation (7.7) can be solved by the variable transform. The substitution $y_r = r\psi_r$ reduces (7.7) to the following form:

$$d^2 y_r / dr^2 = \kappa^2 y \tag{7.10}$$

The differential equation is easily solved, and the solution becomes

$$y_r = A \exp(-\kappa r) + B \exp(\kappa r) \tag{7.11}$$

or

$$\psi_r = A \exp(-\kappa r)/r + B \exp(\kappa r)/r \tag{7.11'}$$

where A and B are the integration constants determined from the physical conditions. Since $\psi_r \to 0$ as $r \to \infty$ we have $B = 0$.

The next problem is to determine another constant A, which can be done using conditions (7.4) and (7.6). In the same way as for (7.7), the following equation is derived from (7.6):

$$\rho_r = -(A\kappa^2 \varepsilon / 4\pi) \times [\exp(-\kappa r)/r] \tag{7.12}$$

where $\psi_r = A \exp(-kr)/r$ is used for ψ_r in (7.6). Introducing ρ_r from (7.12) into (7.4), we have

$$A\kappa^2 \varepsilon \int_{r_0}^{\infty} r \exp(-\kappa r)\, dr = ze \tag{7.13}$$

Integrating (7.13) by parts gives

$$A = ze \exp(\kappa r_0)/[\varepsilon(1 + \kappa r_0)] \tag{7.14}$$

Finally, we obtain the following equation for the electrical potential ψ_r around a central ion whose charge is $z_j e$:

$$\psi_r = z_j e \exp[-\kappa(r - a)]/[\varepsilon r(1 + \kappa a)] \tag{7.15}$$

Here a is used in place of r_0 (the distance between ions at closest approach). The variable a is the sum of the effective radii of ions in solution and is the same for all pairs of ions (a rather bold assumption). Equation (7.25) is the Debye-Hückel equation for dilute electrolyte solutions, and is the fundamental equation for evaluating the activity coefficients of ionic species in solution.

As is clear from the above derivation of (7.15), the electrical potential ψ_r is the sum of the contributions from the central ion and the surrounding ions. These are additive by the principle

$$\psi_r = \psi_r' + \psi_r'' \tag{7.16}$$

where ψ_r' is the contribution of the central ion to the potential and ψ_r'' is the contribution by all of the surrounding ions (the *ion atmosphere*). The potential ψ_r' in the same dielectric continuum is given by

$$\psi_r' = z_j e / \varepsilon r \tag{7.17}$$

Hence, the potential ψ_r'' becomes

$$\psi_r'' = (z_j e / \varepsilon r) \times \{\exp[-\kappa(r - a)]/(1 + \kappa a) - 1\} \tag{7.18}$$

This equation is applicable over the range $r > a$. On the other hand, no other ions enter the spherical region $r < a$, and therefore the potential ψ_a'' remains constant within the sphere. The substitution $r = a$ reduces (7.18) to

$$\psi_a'' = -z_j e \kappa / \varepsilon (1 + \kappa a) = -(z_j e / \varepsilon) \times [1/(a + 1/\kappa)] \tag{7.19}$$

This is the electrical potential imposed on the central ion by all of the surrounding ions when the charge of the central ion is $z_j e$. The central ion then comes to occupy the center of the spherically constant potential due to the ion atmosphere, which results, in turn, from the electrical potential of the central ion. It can be said from (7.19) that the sourrounding ions are distributed over a spherical surface at a distance $(a + 1/\kappa)$ from the center of the sphere, so that their total charge is equal and opposite to the charge of the central ion. The parameter $1/\kappa$, called the *radius of the ion atmosphere*, represents the distance of the ion atmosphere from the central ion at closest approach (Fig. 7.1). Therefore, the free energy change ΔG of the central ion owing to electrical interaction with the surrounding ion atmosphere is obtained by the following evaluation:

$$\Delta G = \int_0^{z_j e} \psi_a'' \, d\rho = -\int_0^{z_j e} [\kappa z_j e / \varepsilon (1 + \kappa a)] \, d(z_j e)$$

$$= -\kappa z_j^2 e^2 / [2\varepsilon (1 + \kappa a)] \tag{7.20}$$

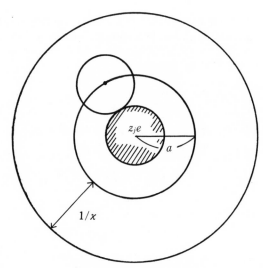

Figure 7.1. Schematic illustration of charge distribution around central ion of charge $z_j e$. $1/\kappa$, radius of ion atmosphere.

sphere of ion atmosphere

where ρ is the charge in the electric field ψ_a''. This free energy is also the excess energy of ideal mixing and contributes to the activity coefficient γ_j of an ion with charge $z_j e$. By analogy with (3.34) and (3.35), the activity coefficient takes the form

$$kT \ln \gamma_j = -\kappa z_j^2 e^2 / [2\varepsilon(1 + \kappa a)] \tag{7.21}$$

On the other hand, κ is related to the ionic strength I

$$\kappa = (8\pi N e^2 / 1000\varepsilon kT)^{1/2} I^{1/2} \tag{7.22}$$

because the parameters in (7.8) have the following relations:

$$n_i^0 = C_i N / 1000 \tag{7.23}$$

$$I = \sum_i C_i z_i^2 / 2 \tag{7.24}$$

where N is Avogadro's number and C_i is a molar concentration. Equation (7.21) is an individual ionic activity coefficient, but it cannot be independently determined by experiment. The activity coefficient that can be determined is the mean activity coefficient $\gamma_\pm$, which is given by

$$\ln \gamma_\pm = -(|z_+ z_-| e^2 / 2\varepsilon kT) \times [\kappa / (1 + \kappa a)] \tag{7.25}$$

for an electrolyte that dissociates in solution to produce ν_+ cations of valency z_+ and ν_- anions of valency z_-, where the following equations are employed:

$$\nu_+ \ln \gamma_+ + \nu_- \ln \gamma_- = (\nu_+ + \nu_-) \ln \gamma_\pm \qquad (7.26)$$

$$\nu_+ z_+ + \nu_- z_- = 0 \qquad (7.27)$$

Finally, the mean activity coefficient in aqueous solution is expressed in terms of the ionic strength I as

$$\log \gamma_\pm = -0.512 |z_+ z_-| \sqrt{I} / (1 + 0.329 \times 10^8 a \sqrt{I}) \qquad (7.28)$$

where $\varepsilon = 78.3$ at 298.15 K and a is the length in centimeters.

7.2. The Diffuse Double Layer

The distribution of ions under an electric potential is determined by their potential energy and their thermal kinetic energy. This is also the case for ions coexisting in solution with colloidal particles. When one colloidal particle is immersed in an ionic solution, the particle is surrounded by an electric double layer. One layer is formed by charges at the surface of the particle, arising either from ions adsorbed out of solution or from dissociated ions of the particle components. The other layer, called the *diffuse layer*, is formed by an excess of oppositely charged ions in the solution adjacent to the charged surface of the particle. The double layer therefore results from an unequal distribution of positive and negative ions at a charged phase boundary (Fig. 7.2). Even though the ions are of finite size, the surface charge is assumed to be homogeneous, and the ions in the solution either are regarded as point charges or are assumed to carry a continuous space charge owing to their time average, resulting in a smooth change of electric potential.

The colloidal particle is assumed to be so large compared to the ions that its surface may be regarded as a plane. The electric potential and the charge density then become a function of the distance l from the charged surface (a flat charged plane). Hence, the Poisson equation (7.2) is written as

$$d^2 \psi_l / d l^2 = -4 \pi \rho_l / \varepsilon \qquad (7.29)$$

The key difference between this presentation and the Debye-Hückel theory is that the electrical surface potential is not always smaller than the

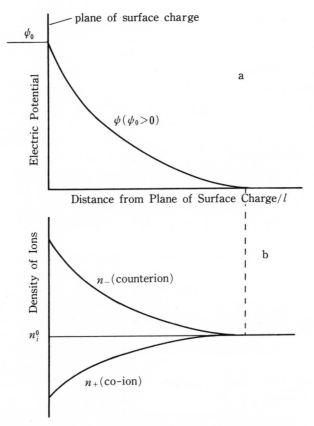

Figure 7.2. Diffuse double layer and ion distribution: (a) electric potential, (b) concentrations of counterions and coions.

kinetic energy. For example, at 25°C the electric potential corresponding to the kinetic energy kT is 25.7 mV. The charge density in the diffuse layer (determined by the equilibrium distribution of every ion) is also expressed by (7.6), but the exponential term in (7.6) cannot be expanded in a linear form as in the Debye–Hückel theory. Let us make the following simplifying assumptions that electrolytes in the solution are symmetric: $z_+ = |z_-| = z$ and $n_+^0 = n_-^0 = n^0$. The Poisson–Boltzmann equation then becomes

$$d^2\psi/dl^2 = -(4\pi ze/\varepsilon) \times (n_+ - n_-) = (8\pi n^0 ze/\varepsilon)\sinh(ze\psi/kT) \quad (7.30)$$

Now, the following substitutions

$$y = ze\psi/kT \quad \text{and} \quad x = \kappa l \quad (7.31)$$

reduce (7.30) to the simplified form

$$d^2y/dx^2 = \sinh y \qquad (7.32)$$

where the parameter κ^2 from (7.8) reads as

$$\kappa^2 = 8\pi n^0 z^2 e^2 / \varepsilon kT \qquad (7.33)$$

Multiplying both sides of (7.32) by $2dy/dx$ and integrating once with respect to x under the boundary conditions, $y = 0$ and $dy/dx = 0$ for $x = \infty$, we obtain

$$dy/dx = -(2\cosh y - 2)^{1/2} = -2\sinh(y/2) \qquad (7.34)$$

The minus sign results from the slope of y, which is assumed to be positive here (Fig. 7.2). After a second integration of (7.34) under the conditions $\psi = \psi_0$ or $y = y_0$ for $x = 0$, we obtain

$$\exp(y/2) = \{\exp(y_0/2) + 1 + [\exp(y_0/2) - 1] \times \exp(-x)\}/$$
$$\{\exp(y_0/2) + 1 - [\exp(y_0/2) - 1] \times \exp(-x)\} \qquad (7.35)$$

Rearranging (7.35) using the parameters of (7.31) and (7.33), we finally obtain

$$\psi = (2kT/ze) \times \ln\{[1 + g\exp(-\kappa l)]/[1 - g\exp(-\kappa l)]\} \qquad (7.36)$$

where

$$g = [\exp(ze\psi_0/2kT) - 1]/[\exp(ze\psi_0/2kT) + 1] \qquad (7.37)$$

The electrical potential ψ decays exponentially against larger l, irrespective of the magnitude of ψ_0. When the electrical potential is much less than 25.7 mV, on the other hand, the Poisson–Boltzmann equation takes the form

$$d^2\psi/dl^2 = \kappa^2\psi \qquad (7.38)$$

Then the solution is the same as for (7.10) and becomes

$$\psi = \psi_0 \exp(-\kappa l) \qquad (7.39)$$

This is the Debye-Hückel approximation. $1/\kappa$ is called the thickness of the double layer and is the distance from the surface to the center of charges in the double layer, as can be seen below.

By analogy with (7.4), the surface charge density σ_0 is written from the electroneutrality condition as

$$\sigma_0 = -\int_0^\infty \rho_l \, dl \qquad (7.40)$$

By introducing (7.29) for ρ into (7.40), there results

$$\sigma_0 = (\varepsilon/4\pi) \times \int_0^\infty (d^2\psi/dl^2) \, dl = -(\varepsilon/4\pi)(d\psi/dl)_{l=0} \qquad (7.41)$$

The surface charge density σ_0 is now found to be proportional to the initial slope of the electric potential. Furthermore, the initial slope can be evaluated by (7.34) to be

$$(d\psi/dl)_{l=0} = -(8\pi n^0 kT/\varepsilon)^{1/2} \times (2\cosh y_0 - 2)^{1/2}$$
$$= -(32\pi n^0 kT/\varepsilon)^{1/2} \times \sinh(y_0/2) \qquad (7.42)$$

and therefore

$$\sigma_0 = (2n^0 \varepsilon kT/\pi)^{1/2} \times \sinh(ze\psi_0/2kT) \qquad (7.43)$$

In cases where the Debye-Hückel approximation is applicable (when $\psi \ll 25.7$ mV), σ_0 becomes simpler

$$\sigma_0 = (\varepsilon\kappa/4\pi)\psi_0 = [\varepsilon/4\pi(1/\kappa)]\psi_0 \qquad (7.44)$$

Now the relation between σ_0 and ψ_0 is just like that of a parallel plate condenser, where the distance $1/k$ between the plates is roughly expressed by

$$1/\kappa \simeq 4.3 \times 10^{-8}/z\sqrt{C} \qquad (\text{cm}) \qquad (7.45)$$

for an aqueous solution at 25°C where $\varepsilon = 78.3$ and C is the molar concentration. Thus, the thickness of the double layer increases with decreasing concentration of ions.

7.3. Potential Energy Due to Electrical Double Layers

When two charged colloidal particles approach each other, the electrical double layers overlap and interact, causing the free energy of the two particles to change. Two ways to evaluate the free energy have been reported: one, by Verwey and Overbeek,[3,4] is based on an electrical energy, and the other, by Derjaguin and Landau,[5,6] is derived from a force working between the two particles. The two give essentially identical results. The present discussion is made from the latter point of view, where the problem is to evaluate the electrical repulsive interaction as a function of the distance between two particles having the same electric charge. As above, the particle is assumed to be large enough that its surface can be regarded as flat.

Figure 7.3 shows two large parallel plates with surface electric potentials ψ_0 spaced $2d$ apart. The solution for (7.32) satisfying the boundary conditions $y = y_d = ze\psi_d/kT$ and $dy/dx = 0$ for $l = d$ is similarly obtained in the form

$$dy/dx = -(2\cosh y - 2\cosh y_d)^{1/2} \tag{7.46}$$

or

$$(d\psi/dl)_{l=0} = -(8\pi n^0 kT/\varepsilon)^{1/2} \times (2\cosh y_0 - 2\cosh y_d)^{1/2} \tag{7.46'}$$

The solution of (7.46) is not given in a closed form, but is available from a table.[7] Let us suppose, as a proper approximation, that the electrical potential y_d at $l = d$ is a sum of the potentials of the two double layers, when the overlap between the double layers is small (the two plates are far apart). In this case, $\kappa d \gg 1$, (7.36) can be transformed to the simpler form

$$y \simeq 2\ln\{1 + 2g\exp(-x)\} \simeq 4g\exp(-x) \tag{7.47}$$

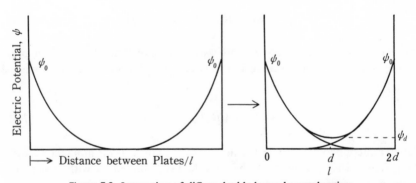

Figure 7.3. Interaction of diffuse double layers by overlapping.

Then we have the following equation for y_d :

$$y_d = 2y = 8g \exp(-\kappa d) \tag{7.48}$$

Next, let us consider a balance of forces impinging on a space charge in a volume element of volume dl, unit area, and thickness dl located in the diffuse layer (Fig. 7.4). The gradient of hydrostatic pressure dP and the force on a space charge ρ under an electric field should balance each other:

$$dP + \rho \, d\psi = 0 \tag{7.49}$$

Introducing (7.29) into (7.49), we have

$$(d/dl)[P - (\varepsilon/8\pi)(d\psi/dl)^2] = 0 \tag{7.50}$$

or

$$P - (\varepsilon/8\pi)(d\psi/dl)^2 = \text{constant} \tag{7.50'}$$

Thus, the difference between the hydrostatic pressure and the electric force f_E turns out to be constant at every point. The physical condition that $P = P_\infty$ for $d\psi/dl = 0$ at the point far from the charged surface leads to

$$P - f_E = P_\infty \tag{7.51}$$

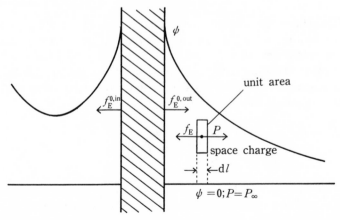

Figure 7.4. Forces in balance imposed upon space charge in volume element (dl) in diffuse layer: f_E, electric force; P, pressure.

and

$$f_E = (\varepsilon/8\pi)(d\psi/dl)^2 \tag{7.52}$$

The next problem is to evaluate the electrostatic force imposed upon the charged surface by an excess opposite charge in the diffuse layer. The electric force expressed by (7.52) presses the space charge toward the charged surface. This force gradually increases from zero to the final value

$$f_E^0 = (\varepsilon/8\pi)\,(d\psi/dl)_0^2 \quad \text{at } l = 0 \tag{7.52'}$$

in the thin liquid layer adjacent to the charged surface. Conversely, the charged surface of the plate experiences an equal force from the thin layer working toward the outer bulk solution, owing to the excess opposite charge in the diffuse layer. As is clear from (7.52'), the value of $[(d\psi/dl)_0]_d$ at the inner plate surface is a function of the plate separation, and decreases with decreasing distance between the plates. The force from this slope works toward the inner bulk solution. Hence, the resultant force pointing outward can be written as

$$f_E^r = (\varepsilon/8\pi) \times \{[(d\psi/dl)_0^2]_{d=\infty} - [(d\psi/dl)_0^2]_{d=d}\} \tag{7.53}$$

Introducing (7.42) and (7.46') into (7.53) leads finally to

$$f_E^r = 2n^0 kT(\cosh y_d - 1) \tag{7.54}$$

We can conclude that uneven charge distribution in the diffuse layers on both sides of a plate leads to an excess force that pushes the plates apart. The same result was also derived by Langmuir,[8] who used the osmotic pressure caused by the excess ions present between the two plates.

The potential energy V_R for the resultant repulsive force is given by the following integration for the two plates:

$$V_R = -2 \int_\infty^d f_E^r \, dd \tag{7.55}$$

In this case, the electrical potential y_d midway between the plates can be small ($y_d < 1$), and then the repulsive force of (7.54) simplifies to

$$f_E^r = n^0 kT y_d^2 = 64 n^0 kT g^2 \exp(-2\kappa d) \tag{7.56}$$

where the potential of (7.48) is used for (7.56). By introducing (7.56) into (7.55) and integrating, we obtain

$$V_R = (64n^0kT/\kappa) \times g^2 \exp(-2\kappa d) \qquad (7.57)$$

This is the potential energy needed to bring the two charged parallel plates from infinite separation to a separation of $2d$ against the repulsive force. According to tabulated values of κd,[7] the range $\kappa d > 2$ satisfies the condition $y_d < 1$. Therefore, (7.57) is valid only for the regon $\kappa d > 2$. Furthermore, the following equations are given for the repulsive potential between spherical particles[4]:

$$V_R(l) = (\varepsilon R\psi_0^2/2) \times \ln[1 + \exp(-\kappa l)] \qquad \text{for } \kappa R \gg 1 \qquad (7.58)$$

$$= [\varepsilon R^2\psi_0^2/(l + 2R)] \times \exp(-\kappa l) \qquad \text{for } \kappa R \ll 1 \qquad (7.58')$$

where the parameters R, r, and l are as shown in Fig. 7.5.

7.4. Potential Energy Due to the van der Waals–London Force

An attractive force between neutral molecules can be explained to some extent by a dipole–dipole interaction when the particles carry a dipole moment. Such an attractive force even exists between nonpolar molecules as a result of the influence of electron motion in one atom on the motion in the other atom. This force was elucidated by London on the basis of wave mechanics.[9] The charge fluctuations in one atom or molecule induce a fluctuating electric dipole in the other atom or molecule. A dipole-induced dipole interaction is thus set up, which in turn leads to an attractive interaction between the atoms or molecules. The attractive potential due to this force is expressed as

$$V_A = -\lambda/r^6 \qquad (7.59)$$

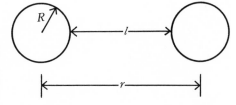

Figure 7.5. Coordinate parameters for two spherical particles: R, radius; l, distance between two surfaces; r, distance between centers of two spheres.

where λ is a constant depending on the properties of atoms or molecules and r is the distance between them. The constant λ is called the *van der Waals-London constant*, and is given by London for two equal atoms as

$$\lambda = 3\alpha^2 h\nu_0/4 \qquad (7.60)$$

where α is polarizability and $h\nu_0$ is the energy corresponding to the chief specific frequency ν_0. Therefore, the attractive energy V_A between two colloidal particles having molecular or atomic density n/cm^3 is obtained by summing the attractive potentials for all pairs formed among them:

$$V_A = -\lambda n^2 \int_{V_1} \int_{V_2} (1/r^6)\, dv_1\, dv_2 \qquad (7.61)$$

where dv_1 and dv_2 are, respectively, the volume elements in the total volumes V_1 and V_2 of colloidal particles 1 and 2.

The expression evaluated by integration will be given below for two cases: one involving two large parallel plates and the other involving two spheres of equal radius. The total attractive force per unit area ($1\ cm^2$) of the plate is finally given as[10]

$$f = (\pi n^2 \lambda/6) \times [1/l^3 + 1/(l+2\delta)^3 - 2/(l+\delta)^3] \qquad (7.62)$$

where the variables l and δ are as shown in Fig. 7.6. The corresponding

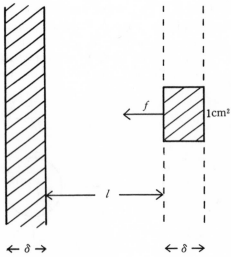

Figure 7.6. Dimensions of two parallel plates and the van der Waals–London force f.

attractive potential V_A is then obtained by integrating with respect to l from ∞ to $2d$:

$$V_A = -\int_{2d}^{\infty} f \, dl$$

$$= -(A/48\pi) \times [1/d^2 + 1/(d + \delta)^2 - 2/(d + \delta/2)^2] \quad (7.63)$$

with the following useful approximations:

$$V_A(2d) = -A\delta^2/32\pi d^4 \quad \text{for } d \gg \delta \quad (7.64)$$

$$= -(A/48\pi) \times (1/d^2 - 7/\delta^2) \quad \text{for } d < \delta \quad (7.64')$$

$$= -A/48\pi d^2 \quad \text{for } d \ll \delta \quad (7.64'')$$

where A is the Hamaker constant:

$$A = \pi^2 n^2 \lambda \quad (7.65)$$

For two spheres of equal radius, the following equation was derived by Hamaker[11]:

$$V_A(r) = -(A/6) \times [2R^2/(r^2 - 4R^2) + 2R^2/r^2 + \ln(1 - 4R^2/r^2)] \quad (7.66)$$

where the variables R, r, and l are as shown in Fig. 7.5. This equation can also be approximated as:

$$V_A(l) = -(A/12)(R/l) \quad \text{for } l \ll R(r \simeq 2R) \quad (7.67)$$

$$= -(16A/9)(R^6/l^6) \quad \text{for } l \gg R \quad (7.67')$$

7.5. Total Potential Energy and the Schulze–Hardy Rule

The total potential energy V_t of two colloidal particles is a sum of V_R and V_A. Whether the two particles coagulate depends on the height of V_{max}—i.e., on the magnitude of the potential barrier of V_t—as shown in Fig. 7.7. When V_{max} is much lager than the thermal kinetic energy of the two particles ($>15\,kT$), coagulation does not take place. Generally, V_{max} depends on V_R rather than V_A, because the contribution of the latter to V_t is always negative. When the surface potential ψ_0 is high and the ionic strength is low, the overlap of two diffuse layers becomes large at a shorter

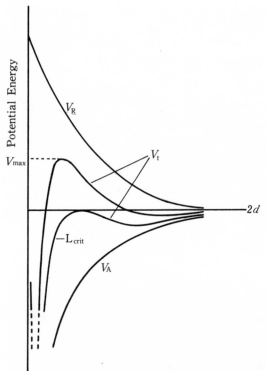

Figure 7.7. Values of V_R, V_A, and V_t as a function of interplate distance $2d$.

interparticle distance, resulting in a high potential barrier and no coagulation. Conversely, at a higher electrolyte concentration the electrical potential decays more rapidly, and V_R is much less at a larger separation. Then V_{max} becomes small enough for coagulation to occur. The electrolyte concentration that brings about coagulation is called the *coagulation concentration* or *critical flocculation concentration* (C_c). This concentration is determined mathematically by such conditions as

$$V_t(2d) = 0 \quad \text{and} \quad dV_t(2d)/dd = 0 \qquad (7.68)$$

which correspond to the L_{crit} line in Fig. 7.7.

For colloidal particles of larger size, Eqs. (7.57) and (7.64″) are applicable for V_R and V_A, respectively. The total energy V_t then becomes

$$V_t = V_R + V_A = (64n^0kT/\kappa) \times g^2 \exp(-2\kappa d) - A/48\pi d^2 = 0 \quad (7.69)$$

and

$$dV_t/dd = -2\kappa V_R - 2V_A/d = 0 \qquad (7.70)$$

The relation $V_R = -V_A$ reduces (7.70) to

$$\kappa d = 1 \tag{7.71}$$

Introducing (7.71) into (7.69) yields

$$(64n^0 kT/\kappa) \times g^2 \exp(-2) = A\kappa^2/48\pi \tag{7.72}$$

Finally, we have the following relations for the concentration C_c:

$$n^0 = 107\varepsilon^3 (kT)^5 g^4/[A^2(ze)^6] \quad \text{or} \quad C_c = 8.28 \times 10^{-25} g^4/A^2 z^6 \tag{7.73}$$

where $T = 298.15$ K, $\varepsilon = 78.3$, and C_c is the molar concentration.

When ψ_0 is large (>200 mV) or $g = 1$, the flocculation concentration C_c is inversely proportional to the sixth power of the valency of the ions. Therefore, for the concentration ratio of C_c one-to-one, two-to-two, and three-to-three types of electrolytes, the following equation is given:

$$C_c^{1-1} : C_c^{2-2} : C_c^{3-3} = 1/1^6 : 1/2^6 : 1/3^6 = 100 : 1.6 : 0.13 \tag{7.74}$$

This is the well-known empirical Schulze–Hardy rule of coagulation concentration, which has been confirmed experimentally in many colloidal systems. The above theory of dispersion and coagulation of colloidal particles, based on the repulsive interaction between two diffuse layers and on the attractive interaction owing to the van der Waals–London force, is called the Derjaguin–Landau–Verwey–Overbeek (D.L.V.O.) theory.

We know that V_A is inversely proportional to d^2, whereas V_R decreases exponentially with the distance $2d$. Thus, at short and long distances of d, V_A becomes larger than V_R, but at intermediate distances the two particles repel each other strongly due to the repulsion of Born and Mayer. Thus, V_t has two minima, as shown in Fig. 7.7: a deep one at a short interparticle distance and a shallow one at a relatively long interparticle distance. Coagulation takes place at the first deep minimum. The secondary minimum plays an important role for plate- or rod-like particles that have a wide interparticle contact area. However, because the second minimum is relatively shallow, the coagulation induced by it is easily broken by an external force. This effect is closely related to rheological phenomena of colloidal suspensions.

Finally, a few remarks are in order concerning the interaction of spherical colloids of radius R. When ψ_0 is relatively small and the radius is much larger than the thickness of the double layer ($R \gg 1/\kappa$), the repulsive potential is given by (7.58). On the other hand, the attractive potential is expressed by (7.67) for $R \gg l$. Thus, the total potential energy becomes

$$V_t = (\varepsilon R\psi_0^2/2) \times \ln [1 + \exp(-\kappa l)] - AR/12l = RV(\psi_0, \kappa, A, l) \tag{7.75}$$

It is evident from (7.75) that V_t is proportional to the radius of colloidal particles. In other words, V_{max} decreases with decreasing radius, and flocculation therefore is easier for small colloidal particles as long as the remaining parameters ψ_0, κ, A, and l remain constant.

The above discussions are based on the electric double layer that originates from the surface potential ψ_0 (Goüy–Chapman model). Strictly speaking, however, ψ_0 is not suitable, because the model includes some assumptions. For example, the typical assumption of point charges leads to absurdly high ion concentrations adjacent to the charged surface for ions of a finite size. In order to take the size of ions into account, Stern[12] placed a plane of the electric potential ψ_δ made of specifically adsorbed counterions at a distance δ from the charged surface of ψ_0, where the counterions are attached to the charged surface strongly enough to overcome their thermal energy. This layer of adsorbed ions is called the *Stern layer*.[13] Thus, more reasonable discussions of the theory should be conducted using ψ_δ instead of ψ_0, where ψ_δ can vary with the kinds of attached ions and with the interparticle distance.[14]

References

1. R. A. Robinson and R. H. Stokes, *Electrolyte Solutions*, Butterworths, London (1959).
2. J. O. Bockris and A. K. N. Reddy, *Modern Electrochemistry*, Plenum Press, New York (1970).
3. E. J. W. Verwey, *J. Phys. Chem. 51*, 631 (1947).
4. E. J. W. Verwey and J. T. G. Overbeek, *Theory of the Stability of Lyophobic Colloid*, Elsevier, Amsterdam (1948).
5. B. V. Derjaguin, *Trans. Faraday Soc. 36*, 203 (1940).
6. B. V. Derjaguin and L. Landau, *Acta Physicochim. URSS 14*, 633 (1941).
7. E. Jahnke and F. Emde, *Tables of Functions*, 2nd ed., p. 124, Teubner, Stuttgart (1933).
8. I. Langmuir, *J. Chem. Phys. 6*, 893 (1938).
9. F. London, *Z. Phys. 63*, 245 (1930).
10. Derivation is given on p. 101 of Ref. 4.
11. H. C. Hamaker, *Physica* (The Hague) *4*, 1058 (1937).
12. O. Stern, *Z. Elektrochem. 30*, 508 (1924).
13. J.T.G. Overbeek, in: *Colloid Science*, Vol. 1, *Irreversible Systems* (H. R. Kruyt, ed.), pp. 132–137, Elsevier, Amsterdam (1952).
14. S. Usui, *J. Colloid I terface Sci. 97*, 247 (1984).

8

Adsorption of Surfactants

8.1. Introduction

One of the most striking physicochemical characteristics of surfactants is their ability to reduce interfacial tension at low concentration. Surface-active agents are defined by this property. Unlike surfactants, inorganic salts increase the interfacial tension. This difference between surfactants and inorganic salts depends on the way the solute is adsorbed at an interface: surfactants show positive adsorption, whereas inorganic salts show negative adsorption. The extent of positive adsorption of a surfactant depends entirely on its chemical structure and on the solvent. This chapter develops the physical meaning of interfacial tension and the relation between interfacial tension change and surfactant concentration in order to elucidate the nature of surface activity.[1-4]

8.2. Surface Tension

Bulk phases contact each other at an *interface*. The region near the interface differs in physicochemical properties from the bulk phase far from the interface. This difference gives rise to *interfacial tension*. In this section, the interfacial tension is discussed first in terms of mechanics and then thermodynamics.

Let us take two homogeneous bulk phases α and β (Fig. 8.1). The interfacial layer between them experiences a normal tension t parallel to the Y axis and varying with Z. The origin A is set up at an arbitrary point in the α phase. The only mechanical force is the pressure in the bulk phases, and the interfacial layer is subject to tension. Consider a rectangular plane

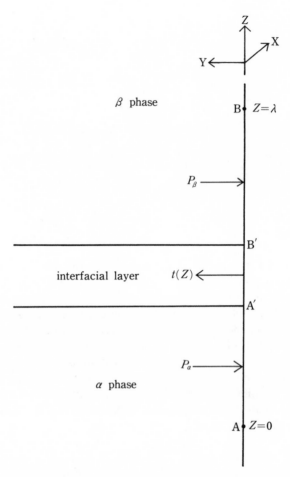

Figure 8.1. Schematic illustration of the interfacial region. P_α, hydrostatic pressure of a phase α up to A′; P_β, hydrostatic pressure of a phase β uniform down to B′; $t(Z)$, tension varying with position Z. (Reproduced with permission of Longman.)

normal to the interface, formed by the movement of the normal AB by one unit length, where point B is located in the bulk phase β. In place of the real situation diagrammed in Fig. 8.1, a simplified, equivalent model is assumed, where the two phases remain uniform up to a geometric surface called the *surface of tension* (Fig. 8.2).[4] The tension t is taken as positive and pressure P as negative. The following two equations result from the assumptions that (1) the total tension across A′B′ of Fig. 8.1 ("reality") is equivalent to the summation of the corresponding interfacial tension and pressure in the model:

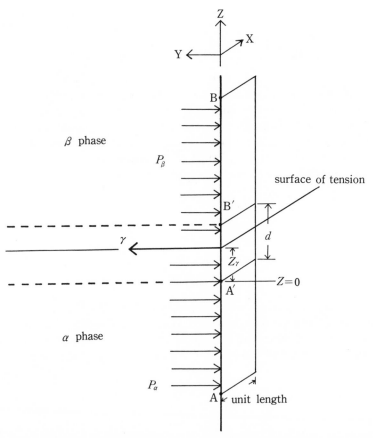

Figure 8.2. Simplified model of interfacial region constructed to place the surface of tension. γ, interfacial tension; d, thickness of surface layer; Z_γ, position of surface of tension on Z axis; other symbols are the same as in Fig. 8.1. (Reproduced with permission of Longman.)

$$\int_0^d t\,\mathrm{d}Z = \gamma - P_\alpha Z_\gamma - P_\beta(d - Z_\gamma) \qquad (8.1)$$

and (2) the moments about A′ are the same in reality and the model

$$\int_0^d tZ\,\mathrm{d}Z = \gamma Z_\gamma - P_\alpha Z_\gamma(Z_\gamma/2) - P_\beta(d - Z_\gamma)[(d + Z_\gamma)/2] \qquad (8.2)$$

The above two equations can be rearranged, respectively, as

$$\gamma = \int_0^{Z_\gamma}(P_\alpha + t)\,\mathrm{d}Z + \int_{Z_\gamma}^d (P_\beta + t)\,\mathrm{d}Z \qquad (8.1')$$

and

$$\gamma Z_\gamma = \int_0^{Z_\gamma} (P_\alpha + t)Z \, \mathrm{d}Z + \int_{Z_\gamma}^{d} (P_\beta + t)Z \, \mathrm{d}Z \qquad (8.2')$$

Thus, the microscopic state of the interfacial layer determines both the value of the surface tension and the location of the surface of tension (Z_γ).

Let us now consider the interfacial tension of a plane interface from the thermodynamic point of view. On the basis of the concept of Guggenheim,[3] the Helmholtz free energy A of the entire system (phases α and β plus the interfacial layer) is expressed by the following equation[5]:

$$\mathrm{d}A = -S \, \mathrm{d}T - P \, \mathrm{d}V + \gamma \, \mathrm{d}a + \sum_i \mu_i \, \mathrm{d}n_i \qquad (8.3)$$

and then

$$\gamma \equiv (\partial A / \partial a)_{T, V, n_i} \qquad (8.4)$$

where γ is an interfacial area. The pressure P and the chemical potentials μ_i remain constant throughout the system because the pressure is the environmental variable and the whole system is in equilibrium. If temperature, pressure, and composition are kept constant, integration of (8.3) leads to

$$\gamma = \left[PV - \left(\sum_i \mu_i n_i - A \right) \right] \Big/ a \qquad (8.5)$$

For interfacial tension in this case, also, the Z axis is normal to the plane interface. The bottom (A) of the whole system is at $Z = 0$ in phase α and the top (B) is at $Z = \lambda$ in phase β (Fig. 8.1). The quantity $c_i(Z)$ is the mean molecular concentration per unit volume of component i at Z, and $A_i(Z)$ is the mean contribution of component i to A at Z. Then, we have

$$V = \int_0^\lambda a \, \mathrm{d}Z \qquad (8.6)$$

$$n_i = \int_0^\lambda a c_i(Z) \, \mathrm{d}Z \qquad (8.7)$$

$$A = \sum_i \int_0^\lambda a c_i(Z) A_i(Z) \, \mathrm{d}Z \qquad (8.8)$$

Substituting Eqs. (8.6), (8.7), and (8.8) into (8.5), we obtain

$$\gamma = \int_0^\lambda [P - P'(Z)]\,dZ \tag{8.9}$$

where

$$P'(Z) = \sum_i c_i(Z)[\mu_i - A_i(Z)] \tag{8.10}$$

In either homogeneous bulk phase, $P'(Z) = P$. Then, the Z range that contributes the surface tension in (8.9) is an inhomogeneous interfacial layer at which a pressure $P'(Z)$ is replaced by a tension opposite in direction to the pressure. Therefore, (8.9) becomes identical with (8.1') and, in addition, the mathematical plane of tension is located by fundamental mechanics, as in (8.2')

$$Z_\gamma = (1/\gamma) \times \int_0^\lambda [P - P'(Z)]Z\,dZ \tag{8.11}$$

The interface is more often curved than plane. This section considers the case in which the surface is spherical. The curvature is assumed to be sufficiently large that (1) edge effects may be neglected and (2) both phases achieve their bulk properties on either side of the interface (Fig. 8.3).

Figure 8.3 shows a spherical cone of solid angle ω where r is the distance from the center of the cone. The cone contains n_i molecules between $r = R_\alpha$ (phase α) and $r = R_\beta$ (phase β) at temperature T, where the constraint is that contours of equal density, etc., lie on spherical surfaces centered on $r = 0$. The equilibrium pressure: in the homogeneous bulk phases are P_α and P_β. By analogy with (8.3) dA for the whole system is

$$dA = -S\,dT - P_\beta \omega R_\beta^2\,dR_\beta + P_\alpha \omega R_\alpha^2\,dR_\alpha + \sigma\,d\omega + \sum_i \mu_i\,dn_i \tag{8.12}$$

where

$$\sigma \equiv (\partial A/\partial \omega)_{T,R_\alpha,R_\beta,n_i} \tag{8.13}$$

Integrating this equation at constant T, R_α, R_β, and composition, we have

$$\sigma\omega = A - \sum_i \mu_i n_i \tag{8.14}$$

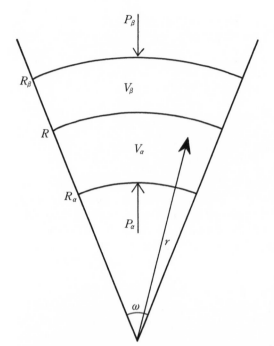

Figure 8.3. Spherical cone including interfacial region. ω, solid angle; r, distance from the center of cone; R_α, position of r for the bottom of the cone container; R_β, position of r for the top of the cone container: R, position of r for the dividing interface. (Reproduced with permission of the American Chemical Society.)

To define the interfacial area $a = \omega R^2$, the dividing surface must be placed somewhere between phases α and β, from the thermodynamic point of view. The two phase volumes V_α and V_β and the interfacial area a can be expressed as follows:

$$V_\alpha = (\omega/3)(R^3 - R_\alpha^3) \qquad (8.15)$$

$$V_\beta = (\omega/3)(R_\beta^3 - R^3) \qquad (8.16)$$

$$a = \omega R^2 \qquad (8.17)$$

Then, their total derivatives become, respectively,

$$dV_\alpha = (d\omega/3)(R^3 - R_\alpha^3) + \omega(R^2\,dR - R_\alpha^2\,dR_\alpha) \qquad (8.18)$$

$$dV_\beta = (d\omega/3)(R_\beta^3 - R^3) + \omega(R_\beta^2\,dR_\beta - R^2\,dR) \qquad (8.19)$$

$$da = R^2\,d\omega + 2\omega R\,dR \qquad (8.20)$$

Solving (8.18), (8.19), and (8.20) for dR_α, dR_β, and $d\omega$ in terms of dV_α, dV_β, da, and dR, substituting these expressions into (8.12), and using (8.14) to eliminate σ, we obtain

$$dA = -S\,dT - P_\alpha\,dV_\alpha - P_\beta\,dV_\beta + \gamma\,da + \xi\,dR + \sum_i \mu_i\,dn_i \quad (8.21)$$

where γ and ξ satisfy the equations

$$\gamma = \left[P_\alpha V_\alpha + P_\beta V_\beta - \left(\sum_i \mu_i n_i - A \right) \right] \Big/ a \quad (8.22)$$

$$\xi = \omega R^2 (P_\alpha - P_\beta) - 2\omega\gamma R \quad (8.23)$$

Equation (8.22) is consistent with the result of integrating (8.21) at constant T, R, an composition—that is, (8.21) applies to a system in complete equilibrium. Therefore, A does not vary with displacement of the interface, and R is not an independent variable. This leads to the cancellation of the $\xi\,dR$ term ($\xi = 0$), and the dividing surface is automatically determined from this condition

$$2\gamma/R = P_\alpha - P_\beta \quad (8.24)$$

Then dA becomes

$$dA = -S\,dT - P_\alpha\,dV_\alpha - P_\beta\,dV_\beta + \gamma\,da + \sum_i \mu_i\,dn_i \quad (8.25)$$

and a very useful expression for surface thermodynamics is obtained from (8.22):

$$a\,d\gamma = -S\,dT + V_\alpha\,dP_\alpha + V_\beta\,dP_\beta - \sum_i \mu_i\,dn_i \quad (8.26)$$

In the plane interface, $P_\alpha = P_\beta = P$ and $V = V_\alpha + V_\beta$. Thus, (8.26) becomes

$$a\,d\gamma = -S\,dT + V\,dP - \sum_i n_i\,d\mu_i \quad (8.27)$$

which is the Gibbs-Duhem equation for an interface.

8.3. Thermodynamics of Adsorption

Let us consider two bulk phases divided by an interface. Figure 8.4 shows the interfacial region, composed of two homogeneous bulk phases α and β and an inhomogeneous interfacial layer. When a mathematical dividing plane (of thickness zero) is placed somewhere in the interfacial region, the total amount in moles n_i of the component is given by the following equation:

$$n_i = n_i^\sigma + n_i^\alpha + n_i^\beta \tag{8.28}$$

where n_i^α and n_i^β are the moles of component i present in phases α and β, respectively, assuming the phases remain homogeneous up to the dividing surface, and n_i^σ is the number of moles adsorbed at the interface. The latter quantity, called the *interfacial excess* (the *surface excess* if the β phase is

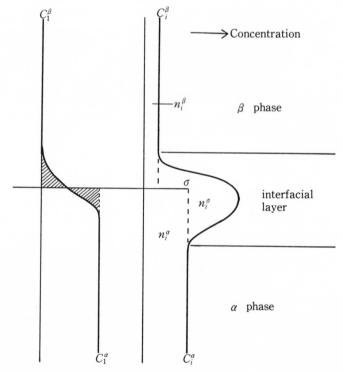

Figure 8.4. Exaggerated illustration of the interfacial region. σ, dividing surface; C_i^α and C_i^β, concentrations of species i in α and β phases, respectively.

gaseous) is equal to the area shown in Fig. 8.4. Thus, the total volume becomes

$$V = V^\alpha + V^\beta \qquad (8.29)$$

and the total mole of component i is written as

$$n_i = n_i^\sigma + Vc_i^\beta + V^\alpha(c_i^\alpha - c_i^\beta) \qquad (8.30)$$

where c_i^α and c_i^β are the concentrations of component i in moles per unit volume in the homogeneous bulk phases α and β, respectively. From (8.30), it is now evident that n_i^σ depends only on V^α or the position of the dividing interface, because the other variables are definite. The adsorption Γ_i of component i is defined by

$$\Gamma_i \equiv n_i^\sigma/a \qquad (8.31)$$

It is very important to select thermodynamic quantities that are invariant with respect to the position of the dividing interface, especially for the discussion on the surface quantities based on experimental results. The volume V^α is expressed by the following equation with respect to component 1 from (8.30):

$$V^\alpha = (a\Gamma_1 - n_1 + Vc_1^\beta)/(c_1^\beta - c_1^\alpha) \qquad (8.32)$$

After introducing (8.32) into (8.30) and rearranging, we obtain the following equation for the relative adsorption of $\Gamma_{i(1)}$ of i with respect to component 1 (by convention, the solvent is designated component 1):

$$\Gamma_{i(1)} \equiv \Gamma_i - \Gamma_1(c_i^\beta - c_i^\alpha)/(c_1^\beta - c_1^\alpha)$$
$$= [n_i - Vc_i^\beta - (n_1 - Vc_1^\beta)(c_i^\beta - c_i^\alpha)/(c_1^\beta - c_1^\alpha)]/a \qquad (8.33)$$

The relative adsorption amount $\Gamma_{i(1)}$ turns out to be invariant with respect to the dividing interface, as is clear from the definite values of the variables in the right-hand side of the last equality, even though Γ_i and Γ_1 depend on the location of the dividing surface.

The next step is to make clear the Gibbs adsorption amount, using the above relative adsorption. Recall (8.27), in which the interfacial tension is a function of $i + 2$ independent variables. However, the Gibbs phase rule permits only i independent variables for two phases including i components. Therefore, the problem is how to reduce the number of intensive variables by two while keeping thermodynamical consistency. The Gibbs–Duhem equations for two homogeneous phases α and β, respectively, are

$$-S^\alpha \, dT + V^\alpha \, dP - \sum_i n_i^\alpha \, d\mu_i = 0 \qquad (8.34)$$

$$-S^\beta \, dT + V^\beta \, dP - \sum_i n_i^\beta \, d\mu_i = 0 \qquad (8.35)$$

Introducing (8.34) and (8.35) into (8.27), we have

$$a \, d\gamma = -S^\sigma \, dT - \sum_i n_i^\sigma \, d\mu_i \qquad (8.36)$$

where the interfacial entropy S^σ

$$S^\sigma = S - S^\alpha - S^\beta \qquad (8.37)$$

and (8.29) are employed. As for (8.37), the interfacial energies are defined respectively as

$$U^\sigma = U - U^\alpha - U^\beta \qquad (8.38)$$

$$A^\sigma = A - A^\alpha - A^\beta \qquad (8.39)$$

$$G^\sigma = A^\sigma - a\gamma = \sum_i \mu_i n_i^\sigma \qquad (8.40)$$

We have now reduced the degrees of freedom by one, employing (8.29) with respect to the volume. On the other hand, from (8.34) and (8.35), the Gibbs–Duhem equations per unit volume for homogeneous bulk phases become the following:

$$dP = s^\alpha \, dT + \sum_i c_i^\alpha \, d\mu_i \qquad (8.41)$$

$$dP = s^\beta \, dT + \sum_i c_i^\beta \, d\mu_i \qquad (8.42)$$

for phases α and β, respectively. From (8.41) and (8.42) $d\mu_1$ becomes

$$d\mu_1 = -[(s^\alpha - s^\beta)/(c_1^\alpha - c_1^\beta)] \, dT - \sum_{i=2} [(c_i^\alpha - c_i^\beta)/(c_1^\alpha - c_1^\beta)] \, d\mu_i \qquad (8.43)$$

Introduction of the above $d\mu_1$ into (8.36) leads to

$$d\gamma = -(1/a)[S^\sigma - n_1^\sigma(s^\alpha - s^\beta)/(c_1^\alpha - c_1^\beta)] \, dT$$
$$- \sum_{i=2} [\Gamma_i - \Gamma_1(c_i^\alpha - c_i^\beta)/(c_1^\alpha - c_1^\beta)] \, d\mu_i \qquad (8.44)$$

This is the *Gibbs-Duhem equation for surface adsorption*. Now γ can be expressed in terms of i independent intensive variables, which is consistent with the phase rule. The important point is that the coefficient of the second term is the relative adsorption of component i with respect to component 1. Then, we have

$$(\partial\gamma/\partial T)_{\mu_i} = -[S^\sigma - n_i^\sigma(s^\alpha - s^\beta)/(c_1^\alpha - c_1^\beta)]/a \qquad (8.45)$$

$$(\partial\gamma/\partial\mu_i)_{T,\mu_{j\neq i}} = \Gamma_{i(1)} = \Gamma_i - \Gamma_1(c_i^\alpha - c_i^\beta)/(c_1^\alpha - c_1^\beta) \qquad (8.46)$$

It should be stressed that the above derivatives are independent of the placement of the dividing surface. When the dividing surface is placed such that $\Gamma_1 = 0$, $\Gamma_{i(1)}$ becomes equal to Γ_i, and the dividing surface is called the Gibbs dividing surface.

The above procedure for reducing the intensive variables by two is too conventional, even though the result is consistent with the phase rule. In fact, we cannot use (8.44) to discuss or measure the effects of pressure on surface phenomena because it has no $V\,dP$ term; in reality, interfacial tension is a function of pressure.

The following discussion is devoted to the mathematical procedure for decreasing the number of degrees of freedom by two from (8.27) for a plane interface.[6] The interfacial tension γ is a function of $i + 2$ variables (T, P, μ_i) by this equation. However, not all of these are independent; two of them can be expressed by the remaining i variables from (8.41) and (8.42), resulting in i degrees of freedom according to the Gibb s phase rule. The elegant way to eliminate two from $i + 2$ variables is to introduce two parameters, λ^α and λ^β. Multiplying (8.41) by λ^α and (8.42) by λ^β, adding them to (8.27), and dividing by a, we obtain

$$d\gamma = -(s - \lambda^\alpha s^\alpha - \lambda^\beta s^\beta)\,dT + (\lambda - \lambda^\alpha - \lambda^\beta)\,dP$$
$$- \sum_{i=1} (n_i^s - \lambda^\alpha c_i^\alpha - \lambda^\beta c_i^\beta)\,d\mu_i \qquad (8.47)$$

where $s = S/a$, $\lambda = V/a$, and $n_i^s = n_i/a$. Any two differentials on the right-hand side of (8.47) are eliminated simply by choosing λ^α and λ^β so that their coefficients become zero. Alternatively, these differentials can be eliminated by expressing any two differentials by the remaining i differentials with use of (8.41) and (8.42) and introducing them into (8.27). Defining s^s, τ^d, and Γ_i, respectively, as

$$s^s = s - \lambda^\alpha s^\alpha - \lambda^\beta s^\beta \qquad (8.48)$$

$$\tau^d = \lambda - \lambda^\alpha - \lambda^\beta \tag{8.49}$$

$$\Gamma_i = n_i^s - \lambda^\alpha c_i^\alpha - \lambda^\beta c_i^\beta \tag{8.50}$$

we obtain the following equation:

$$d\gamma = -s^s\, dT + \tau^d\, dP - \sum_{i=1} \Gamma_i\, d\mu_i \tag{8.51}$$

The point is that any two independent variables can be eliminated simply by making their coefficients equal to zero.

The Gibbs adsorption equation corresponds to selecting τ^d and Γ_1 as the coefficients of the two variables:

$$\lambda - \lambda^\alpha - \lambda^\beta = 0$$

$$n_1^s - \lambda^\alpha c_1^\alpha - \lambda^\beta c_1^\beta = 0 \tag{8.52}$$

Then, λ^α and λ^β become respectively:

$$\lambda^\alpha = (\lambda c_1^\beta - n_1^s)/(c_1^\beta - c_1^\alpha); \qquad \lambda^\beta = (n_1^s - \lambda c_1^\alpha)/(c_1^\beta - c_1^\alpha) \tag{8.52'}$$

and Γ_i becomes

$$\Gamma_i = n_i^s - \lambda c_i^\beta - (n_1^s - \lambda c_1^\beta)(c_i^\beta - c_i^\alpha)/(c_1^\beta - c_1^\alpha) \tag{8.53}$$

which is the same as Eq. (8.33) for the relative adsorption of i with respect to component 1.

Surface tension is ususally measured in the presence of air (component 2), and the effects of concentration of another component, temperature, and pressure on the surface tension are examined to determine respectively the relative adsorption, the entropy change, and the volume change upon adsorption. For this examination, the two coefficients to be eliminated are those of the solvent (component 1), and air:

$$n_1^s - \lambda^\alpha c_1^\alpha - \lambda^\beta c_1^\beta = 0; \qquad n_2^s - \lambda^\alpha c_2^\alpha - \lambda^\beta c_2^\beta = 0 \tag{8.54}$$

Thus, (8.51) reduces to

$$d\gamma = -s^s\, dT + \tau^d\, dP - \sum_{i=3} \Gamma_i\, d\mu_i \tag{8.55}$$

and the derivatives of γ with respect to temperature, pressure, and chemical potential respectively become

$$(\partial\gamma/\partial T)_{P,\mu_i} = -s^s \tag{8.56}$$

$$(\partial\gamma/\partial P)_{T,\mu_i} = \tau^{\mathrm{d}} \tag{8.57}$$

$$(\partial\gamma/\partial\mu_i)_{T,P,\mu_{j\neq i}} = -\Gamma_i \tag{8.58}$$

where λ^α and λ^β have respectively the following equations:

$$\lambda^\alpha = (n_1^{\mathrm{s}}c_2^\beta - n_2^{\mathrm{s}}c_1^\beta)/(c_1^\alpha c_2^\beta - c_1^\beta c_2^\alpha)$$

$$\lambda^\beta = (n_2^{\mathrm{s}}c_1^\alpha - n_1^{\mathrm{s}}c_2^\alpha)/(c_1^\alpha c_2^\beta - c_1^\beta c_2^\alpha) \tag{8.54'}$$

As a result, Γ_i becomes invariant with respect to the dividing plane positions λ^α and λ^β

$$\Gamma_i = n_i^{\mathrm{s}} - c_i^\alpha(n_1^{\mathrm{s}}c_2^\beta - n_2^{\mathrm{s}}c_1^\beta)/(c_1^\alpha c_2^\beta - c_1^\beta c_2^\alpha)$$

$$-c_i^\beta(n_2^{\mathrm{s}}c_1^\alpha - n_1^{\mathrm{s}}c_2^\alpha)/(c_1^\alpha c_2^\beta - c_1^\beta c_2^\alpha) \tag{8.59}$$

Figure 8.5 illustrates schematically λ^α, λ^β, τ^{d}, and Γ_i.[7] The slope against pressure of an interfacial tension between two phases of only two components indicates the dimensions of the interfacial range. The question is whether τ^{d} is always positive and what Γ_i means when τ^{d} is negative for a three-component system. Fortunately, however, τ^{d} has been positive in

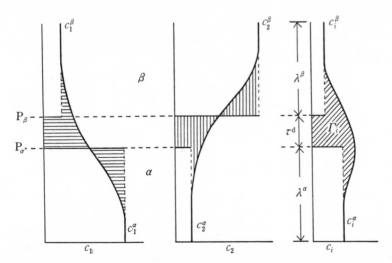

Figure 8.5. Schematic illustration of the two dividing planes, P_α and P_β, and the relative adsorption, Γ_i. (Reproduced with permission of the American Chemical Society.)

many experimental results, judging from the positive slope of the plots of interfacial tension against pressure. Figure 8.6 shows an example of the effect of pressure on the interfacial tension between water and hexane, where tetradecanol as a solute is initially dissolved in hexane. The τ^d value is much less than the molecular size. In this sense, the intuitive omission of the $V\,dP$ term from the Gibbs adsorption equation seems highly reasonable.

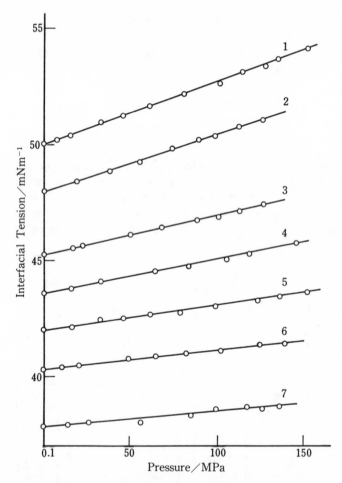

Figure 8.6. Pressure dependence of interfacial tension between aqueous and hexane phases at 303.15 K and at a constant mole fraction x of tetradecanol in a hexane phase. 1, $x = 0$; 2, $x = 1.83 \times 10^{-4}$; 3, $x = 5.29 \times 10^{-4}$; 4, $x = 8.03 \times 10^{-4}$; 5, $x = 1.16 \times 10^{-3}$; 6, $x = 1.56 \times 10^{-3}$; 7, $x = 2.41 \times 10^{-3}$.[8] (Reproduced with permission of Academic Press.)

8.4. Adsorption from Surfactant Solutions

The surfactants to be considered first are simple ionic surfactants, one-to-one electrolytes that dissociate completely. This discussion can easily be extended to nonionic surfactants. The starting equation is Eq. (8.55) originating from Hansen's two dividing surfaces

$$d\gamma = -s^s \, dT + \tau^d \, dP - \Gamma_+ \, d\mu_+ - \Gamma_- \, d\mu_- \tag{8.60}$$

where the subscripts $+$ and $-$ indicate positively and negatively charged dissociated ions. There are three degrees of freedom for three components and two phases, and the most convenient intensive variables are temperature, pressure, and solute concentration. Note only interfacial tension but also chemical potentials are a function of these three variables:

$$d\mu_i = -s_i \, dT + v_i \, dP + (\partial\mu_i/\partial C_i)_{T,P} \, dC_i \tag{8.61}$$

where s_i and v_i are respectively the partial molar entropy and volume of component i. Thus,

$$-(\partial\gamma/\partial T)_{P,C_i} = s^s - \Gamma_+ s_+ - \Gamma_- s_- \tag{8.62}$$

$$(\partial\gamma/\partial P)_{T,C_i} = \tau^d - \Gamma_+ v_+ - \Gamma_- v_- \tag{8.63}$$

$$-(\partial\gamma/\partial C)_{T,P} = \Gamma_+(\partial\mu_+/\partial C)_{T,P} + \Gamma_-(\partial\mu_-/\partial C)_{T,P} \tag{8.64}$$

where C is total surfactant concentration.

The difference in the three derivatives between composition and the chemical potential of the subscript to be kept constant are now clear from (8.56) through (8.58) and (8.62) through (8.64). Electroneutrality must hold for the whole system, and therefore the condition as to Γ_i becomes

$$\Gamma_+ = \Gamma_-(\equiv\Gamma) \tag{8.65}$$

owing to the electroneutrality of homogeneous bulk phases: $\sum z_i n_i = 0$ from (8.28). On the other hand, the concentrations C_+ and C_- are equal to the surfactant concentration C from complete dissociation. Then the relative adsorption is expressed as

$$\Gamma = -(1/2RT)(\partial\gamma/\partial \ln C)_{T,P}/[1 + (\partial \ln f_\pm/\partial \ln C)_{T,P}] \tag{8.66}$$

where the equality of $f_\pm^2 = f_+ \times f_-$ is employed for the respective activity coefficients f_+ and f_- (γ_+ and γ_- in other chapters).

In the case of ionic surfactants that dissociate partially, an undissociated species u appears in the above equalities

$$d\gamma = -s^s \, dT + \tau^d \, dP - \Gamma_+ \, d\mu_+ - \Gamma_- \, d\mu_- - \Gamma_u \, d\mu_u \qquad (8.67)$$

$$\Gamma = \Gamma_u + \Gamma_+ \quad \text{or} \quad \Gamma = \Gamma_u + \Gamma_- \qquad (8.68)$$

and

$$C_+ = C_- = \alpha C \quad \text{and} \quad C_u = (1 - \alpha)C \qquad (8.69)$$

where C is the concentration of the dissociable surfactant and α is the degree of dissociation. From the dissociation equilibrium $\mu_u = \mu_+ + \mu_-$, the partial differentiation of γ with respect to the concentration becomes

$$(\partial\gamma/\partial C)_{T,P} = -(RT\Gamma/C)$$
$$\times \{1 + (\partial \ln f_u/\partial \ln C)_{T,P} + [\partial \ln(1 - \alpha)/\partial \ln C]_{T,P}\} \qquad (8.70)$$

or

$$(\partial\gamma/\partial C)_{T,P} = -(2RT\Gamma/C)$$
$$\times [1 + (\partial \ln f_\pm/\partial \ln C)_{T,P} + (\partial \ln \alpha/\partial \ln C)_{T,P}] \qquad (8.71)$$

where f_u is an activity coefficient of undissociated surfactant. From the above equations the following two limiting cases arise. (1) When a surfactant solute does not dissociate at all ($\alpha = 0$; nonionic surfactant), Γ become from (8.70):

$$\Gamma = -(1/RT)(\partial\gamma/\partial \ln C)_{T,P}/[1 + (\partial \ln f_u/\partial \ln C)_{R,P}] \qquad (8.72)$$

(2) When a solute dissociates completely ($\alpha = 1$), (8.71) becomes the same as (8.66). For extension of the above treatment to multicomponents of solutes, the reader should refer to the literature.[4]

References

1. J. L. Molliet and B. Collie, *Surface Activity*, Van Nostrand, Princeton, N.J. (1950).
2. L. I. Osipow, *Surface Chemistry: Theory and Industrial Applications*, Reinhold, New York (1962).

3. E. A. Guggenheim, *Thermodynamics*, Elsevier/North-Holland, Amsterdam (1977).
4. R. Defay, I. Prigogine, A. Bellemans, and D. H. Everett, *Surface Tension and Adsorption*, Longmans, London (1966).
5. T. L. Hill, *J. Phys. Chem.* **56**, 526 (1952).
6. R. S. Hansen, *J. Phys. Chem.* **66**, 410 (1962).
7. K. Motomura and M. Aratono, *Langmuir* **3**, 304 (1987).
8. K. Motomura, N. Matsubayashi, M. Aratono, and R. Matuura, *J. Colloid Interface Sci.* **64**, 356 (1978).

9

Solubilization

9.1. Introduction

It has long been known that the aqueous solubility of sparingly soluble or insoluble substances can be increased by adding an appropriate third component. Systematic studies using surfactants led to this phenomenon being called *solubilization*.[1,2] Solubilization plays a very important role in industrial and biological processes. McBain and Hutchinson defined solubilization as "a particular mode of bringing into solution substances that are otherwise insoluble in a given medium, involving the previous presence of a colloidal solution whose particles take up and incorporate within or upon themselves the otherwise insoluble material."[1] This definition is too narrow, because the increase in solubility is not always caused by direct introduction of colloidal particles into the system. More often, the enhanced solubility of the solubilizates as colloidal particles is due to the presence of a third component. Therefore, the term *solubilization* has come to have the following very broad definition: "the preparation of a thermodynamically stable isotropic solution of a substance normally insoluble or very slightly soluble in a given solvent by the introduction of an additional amphiphilic component or components."[2]

Numerous applications of chemical engineering—for example, the dissolution of drugs into aqueous solution and their transport through the body, the preparation of agricultural chemical solutions, and the recovery of oil—depend on solubilization by suitable surfactants. In addition, studies of the physical chemistry of bile acids and bile salts, on one hand, and of their physiological function as solubilizers, on the other hand, make it clear that the behavior of bile salts *in vitro* and their functions *in vivo* are closely related.[3] Solubilization will be increasingly important in the future.

This chapter focuses on the study of solubilizates in micellar colloids and develops a mode consistent with thermodynamics. Furthermore, the

Surfactant

solubilizate = drug, substance to be solubilized

solvent

distribution of solubilizates among micelles will be explicated both from thermodynamics and mathematically.

9.2. Phase Rule of Solubilization

Much of the published work on solubilization is on the phase-separation model of the micelle. Accordingly, solubilization has been treated as a partitioning of solubilizate molecules between a micellar phase and the intermicellar bulk phase.[4-10] A few papers are based on the mass-action approach,[11-14] and theoretical discussions from this position have also appeared.[15-17] Unfortunately, papers discussing solubilization from the standpoint of the Gibbs phase rule are very few.[16-21] This section examine solubilization in terms of the phase rule.

If the micelles are regarded as a phase, then adding an excess solubilizate phase means there are three phases (the third is the intermicellar bulk phase). The total number of components is three (solvent, surfactant, and solubilizate), so the presence of three phases makes the system divariant. That would mean that surfactant concentration would be constant at constant temperature and pressure—but, in fact, the maximum additive concentration (MAC) changes with total surfactant concentration. Even if it were postulated that the increase in the MAC with surfactant concentration above the CMC is due to an increase in the total micellar phase, the concentration of solubilizate in the micellar phase should still remain constant, because the concentration is an intensive property of the system and is therefore homogeneous throughout the micellar phase.

If, on the other hand, the micelles are regarded as a phase and the system does not contain an excess solubilizate phase, there are three degrees of freedom. The surfactant concentration is then a unique variable that determines every intensive property of the system at constant temperature and pressure. In other words, the solubilizate monomer concentration in the intermicellar bulk phase (and therefore also in the micellar phase) is set automatically by the surfactant concentration, irrespective of the total solubilizate concentration in the system. This is not only totally incorrect as theory but is contrary to the experimental evidence that the concentration of solubilizates is determined only by the amount added to the system. Clearly, the phase-separation model of micelles and the partition model of solubilization disagree with reality. This contradiction is easily solved by treating the micelles as a chemical species, as shown in the following section.

9.3. Thermodynamics of Solubilization

Micelle formation is well expressed by the following association equilibrium between surfactant monomers S and micelles M[22]:

$$nS \underset{}{\overset{K_n}{\rightleftharpoons}} M \qquad (9.1)$$

where K_n is the equilibrium constant of micelle formation, and where it is assumed that micelles of aggregation number n are monodisperse in the absence of soubilizate. This assumption avoids the difficulties arising from their actual polydispersity, which will be dealt with later. The ideality of the chemical species in solution is assumed because of their low concentrations. The stepwise association equilibria between micelles and solubilizates R are

$$M + R \underset{}{\overset{\bar{K}_1}{\rightleftharpoons}} MR_1 \qquad K_1 = \frac{[MR_1]}{[M][R]}$$

$$MR_1 + R \underset{}{\overset{\bar{K}_2}{\rightleftharpoons}} MR_2$$

$$\cdots$$

$$MR_{m-1} + R \underset{}{\overset{\bar{K}_m}{\rightleftharpoons}} MR_m \qquad (9.2)$$

where MR_i denotes the micelles associated with i molecules of solubilizate, $\bar{K}_i$ is the stepwise association constant between MR_{i-1} and a monomer molecule of solubilizate, and m is an arbitrary number. The total number of components in this system is $m + 4$, including solvent molecules (solvent, S, R, M, $MR_1, \ldots, MR_m$), and the number of phases is one (micellar solution phase).[18] The $m + 1$ equilibrium equations for the micellar system reduce the number of degrees of freedom by $m + 1$, resulting in four degrees of freedom. Hence, at constant temperature and pressure, two other intensive variables can be selected to prescribe the thermodynamic system. Three sets of combinations of intensive variables are examined below to derive other intensive properties.

1. Monomer concentrations of surfactants, [S], and solubilizate, [R]. From monomer–micelle (9.1) and micelle–solubilizate (9.2) equilibria, the micelle concentrations without and with solubilizates are given respectively by the following equations:

$$[M] = K_n[S]^n \qquad (9.3)$$

$$[MR_i] = K_n \left(\prod_{j=1}^{i} \bar{K}_j \right) [S]^n [R]^i \qquad (9.4)$$

where $[MR_i]$ is the concentration of micelles associated with i solubilizate molecules. Thus, the concentration of any species can be determined by these two variables, if the values of K_n and $\bar{K}_j$ are available. The ratio of $[MR_i]$ to the total micellar concentration $\sum_i [MR_i]$, and the average number of solubilizates per micelle, are given by

$$P([MR_i]) = [MR_i]\Big/ \sum_{i=0}^{m} [MR_i]$$

$$= \left(\prod_{j=1}^{i} \bar{K}_j\right)[R]^i \Big/ \left[1 + \sum_{i=1}^{m}\left(\prod_{j=1}^{i}\bar{K}_j\right)[R]^i\right] \qquad (9.5)$$

$$\bar{R} = \sum_{i=0}^{m} i[MR_i]\Big/ \sum_{i=0}^{m} [MR_i] = ([R_t] - [R])/[M_t]$$

$$= \sum_{i=1}^{m} i\left(\prod_{j=1}^{i}\bar{K}_j\right)[R]^i \Big/ \left[1 + \sum_{i=1}^{m}\left(\prod_{j=1}^{i}\bar{K}_j\right)[R]^i\right] \qquad (9.6)$$

2. Total equivalent concentrations of surfactant, $[S_t]$, and solubilizate, $[R_t]$. This is the most commonly used set of variables. The total equivalent concentration of surfactant is

$$[S_t] = [S] + n\left(\sum_{i=0}^{m} [MR_i]\right)$$

$$= [S] + nK_n[S]^n\left[1 + \sum_{i=1}^{m}\left(\prod_{j=1}^{i}\bar{K}_j\right)[R]^i\right] = f([S],[R]) \qquad (9.7)$$

Hence, the total equivalent concentration can also be expressed by the function of $[S]$ and $[R]$. On the other hand, the total equivalent concentration of solubilizate is

$$[R_t] = [R] + \sum_{i=1}^{m} i[MR_i]$$

$$= [R] + K_n[S]^n \sum_{i=1}^{m} i\left(\prod_{j=1}^{i}\bar{K}_j\right)[R]^i = g([S],[R]) \qquad (9.8)$$

[R_t] is also a function of [S] and [R]. Therefore, when [S_t] and [R_t] are selected as two independent thermodynamic variables, [S] and [R] can be obtained as the solution of two simultaneous equations, f and g. Thus, the concentration of MR_i species can be calculated in the same way as in the first case. In this case, too, the distribution of solubilizates among micelles depends on K_n and $\bar{K}_j$ values.

3. Total micelle concentration, [M_t], and average number of solubilizates per micelle, $\bar{R}$. The total micelle concentration is given by

$$[M_t] = K_n[S]^n \left[1 + \sum_{i=1}^{m} \left(\prod_{j=1}^{i} \bar{K}_j \right)[R]^i \right] \qquad (9.9)$$

The average number of solubilizates per micelle, $\bar{R}$, was given by (9.6). By rearranging (9.6), we obtain

$$\bar{R} + \sum_{i=1}^{m} (\bar{R} - i)\left(\prod_{j=1}^{i} \bar{K}_j \right)[R]^i = 0 \qquad (9.10)$$

The solution of (9.10) should give the [R] value in an association equilibrium. By substituting the [R] value into (9.9), one can obtain the equilibrium concentration of monomeric surfactant [S]. Consequently, any intensive properties can be derived in the same way as before from this combination of variables.

Although the three combinations of variables given above are the most commonly used, others are possible; the suitability of a combination depends on the thermodynamic system concerned. As is clear from the above discussion, the factors that specify the thermodynamic system are K_n and $\bar{K}_j$. The $\bar{K}_j$ value, in particular, is an index of interaction between micelles and solubilizates. These values are unconditionally determined by the combination of solubilizate and surfactant.

A brief comment is in order on the maximum additive concentration. The above discussion pertains when the solubilizate concentration is less than its solubility. When solubility is exceeded, an excess solubilizate phase appears in the system, which then has two phases and is trivariant. In this case, the system is fixed by specifying the total surfactant concentration at constant temperature and pressure. This prediction from the phase rule is supported by the observed changes of the phenothiazine MAC with the concentration of sodium dodecyl sulfate (Fig. 9.1).[19]

The change in behavior caused by the appearance of an excess solubilizate phase originates from the fact that the concentration of monomeric

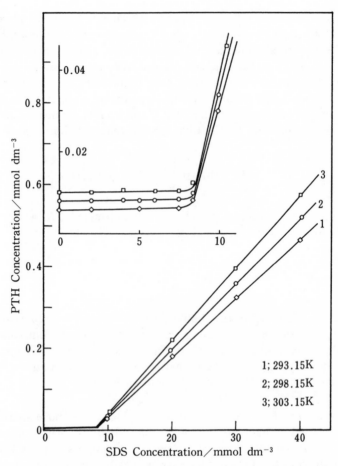

Figure 9.1. Change of the MAC of phenothiazine (PTH) with concentration of sodium dodecyl sulfate (SDS). (Reproduced with permission of the American Chemical Society.)

solubilizate is automatically determined by its solubility at a specified temperature and pressure, and the number of degrees of freedom decreases by one when an excess phase is present. The thermodynamic approach that regards micelles as chemical species perfectly elucidates the solubilization in micellar solutions. An important point, as is clear from the above equation, is that a solubilization system is specified by both the association constants and the monomeric solubilizate concentration. That is, higher association constants do not always lead to higher solubilization. Solubilizates of low monomeric solubility have a smaller total equivalent concentration in spite of their higher association constants (Fig. 9.2).[5,20,23,24]

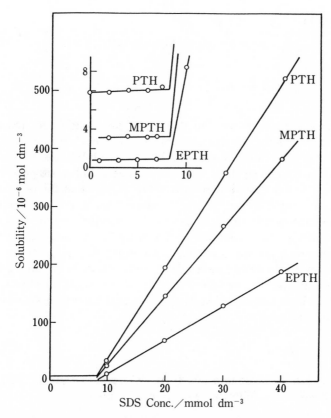

Figure 9.2. Changes in the MAC of phenothiazine (PTH), 10-methylphenothiazine (MPTH), and 10-ethylphenothiazine (EPTH) with concentration of sodium dodecyl sulfate (SDS) at 298.15 K.[20] (Reproduced with permission of the American Chemical Society.)

9.4. Distribution of Solubilizate Molecules among Micelles

If the concentration of solubilizates incorporates into micelles is close to or higher than the micellar concentration, we have to think about the distribution of solubilizates among micelles. As with solubilization, the distribution of solubilizates among micelles can be handled successfully only by rearding micelles as chemical species.[18,19] The equations in Section 9.3 clearly indicate that the association constants between micelles and solubilizates determine the distribution of solubilizates among micelles. The stepwise association constants $\bar{K}_j$ and the concentration of MR_i are given respectively as

$$\bar{K}_j = \vec{k}_{j-1}/\overleftarrow{k}_j = [MR_j]/([MR_{j-1}][R])$$ (9.11)

$$[MR_i] = K_n\left(\prod_{j=1}^{i} \bar{K}_j\right)[S]^n[R]^i$$ (9.4)

The next step is to develop the above equations under reasonable assumptions. When the concentration of solubilizate is less than a few times the micellar concentration, incorporation of the solubilizates into micelles can be assumed to be so slight as not to change the intrinsic properties of the micelles. In this case, it is reasonable to assume that the rate constant value of $\overleftarrow{k}_j$ is j times as large as the $\overleftarrow{k}_1$ value. That is, the probability of a solubilizate molecule escaping from a mother micelle containing j solubilizate molecules is j times the probability of the molecule escaping from a micelle containing just one solubilizate molecule ($\overleftarrow{k}_j = j\overleftarrow{k}_1$). Moreover, the probability of a solubilizate molecule entering a micelle remains the same regardless of the number of solubilizate molecules in the micelle ($\vec{k}_j = \vec{k}_0$). We therefore have the following equation as to the stepwise association of solubilization constants:

$$\bar{K}_j = \bar{K}_1/j$$ (9.12)

If Eq. (9.12) is introduced into Eqs. (9.4), (9.8), and (9.9), then $[MR_i]$, $[R_t]$, and $[M_t]$ become the following equations with infinite m:

$$[MR_i] = \bar{K}_1^i[R]^i[M]/i!$$ (9.13)

$$[R_t] = [R] + \bar{K}_1[R][M] \exp(\bar{K}_1[R])$$ (9.14)

$$[M_t] = [M] \exp(\bar{K}_1[R])$$ (9.15)

Hence, the average number of solubilizate molecules per micelle, $\bar{R}$, is given by

$$\bar{R} = ([R_t] - [R])/[M_t] = \bar{K}_1[R]$$ (9.16)

Thus, the probability that a micelle is associated with i solubilizates can be written from Eq. (9.5) as

$$P(i) = [MR_i]/[M_t] = \bar{R}^i \exp(-\bar{R})/i!$$ (9.17)

This expression is exactly the same as the Poisson distribution. Equation (9.17) is derived by summing m to infinity.

As m becomes high, the present solubilization model ceases to fit the physical state of the micelles, as is clear from the above assumption. However, when the $\bar{R}$ values are low, an important portion of the summation ranges up to only twice $\bar{R}$.[19,25]

The above discussion is based on the thermodynamics of equilibrium associations and on the reasonable assumptions. However, the Poisson equation can also be derived strictly mathematically. Consider a random distribution of r balls in q cells, where both of them are independent and indistinguishable. The probability $P(i)$ that a specified cell contains exactly i balls is given in the form[26]

$$P(i) = \binom{r}{i}(1/q)^i(1 - 1/q)^{r-i} \qquad (9.18)$$

where r and q are on the order of Avogadro's number and i is very small in comparison (less than 10 in most cases). This equation is a special case of the so-called binomial distribution. $P(i)$ can be rearranged as

$$P(i) = (1/i!)(r/q)^i(1 - 1/r) \cdots [1 - (i - 1)/r](1 - 1/q)^{r-i} \qquad (9.19)$$

Using the following approximation equation, which is reasonable for the present discussion,

$$1 - 1/q = \exp(-1/q) \qquad (9.20)$$

we obtain

$$P(i) = (1/i!)\bar{R}^i \exp(-\bar{R})(1 - 1/r) \cdots [1 - (i - 1)/r](1 - 1/q)^{-i} \qquad (9.21)$$

where $\bar{R}$ is the average number of balls per cell, r/q. Because of the conditions as to r, q, and i values, $P(i)$ becomes

$$P(i) = \bar{R}^i \exp(-\bar{R})/i! \qquad (9.17)$$

which is the Poisson distribution, where $r = [R_t] - [R]$ and $q = [M_t]$. The important point here resides in the condition that both balls and cells are independent and indistinguishable, which is possible only when the solubilization is so small that the properties of the micelles do not change.

The Poisson distribution of solubilizates among micelles has been examined photochemically (see Chapter 12), and appears to be correct for a small amount of solubilization. For solubilizations to which the Poisson distribution can be applied, the following useful expression can be derived from (9.14) and (9.15):

$$([R_t] - [R])/[R] = \bar{K}_1[M_t] \qquad (9.22)$$

The K_1 value (the first stepwise association constant between a solubilizate and a vacant micelle) can serve as the normalized interaction parameter between the solubilizate and the micelle. When the results of Fig. 9.2 are plotted, again obeying Eq. (9.22), we obtain the plots shown in Fig. 9.3. The first stepwise association constants are found to increase with increasing hydrophobicity of the solubilizate.

We now address the effects of a polydisperse micellar aggregation number of the thermodynamic expressions of solubilization. One strong effect is variation of the association constant $\bar{K}_j$. We define a new association constant $^n\bar{K}_j$, which is $\bar{K}_j$ of the micelles of aggregation number n. Suppose the polydispersity of the aggregation number of micelles ranges from α to β. The concentration of micelles of aggregation number n with no solubilizate association becomes

$$[M_n] = K_n[S]^n \tag{9.3'}$$

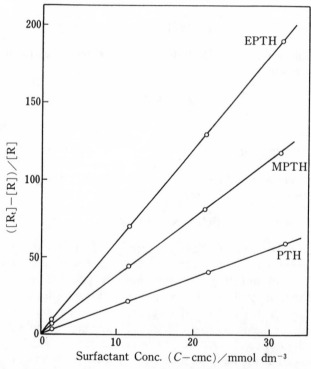

Figure 9.3. Plots of $[R_t] - [R]/[R]$ of PTH, MPTH, and EPTH against micellar concentration $(C - CMC)$ of sodium dodecyl sulfate for determination of $\bar{K}_1$ values at 298.15 K.[20] (Reproduced with permission of the American Chemical Society.)

Summing up $[M_n R_i]$ over the aggregation number from α to β, we obtain for the concentration of micelles associated with i solubilizate molecules

$$[MR_i] = \sum_{n=\alpha}^{\beta} [M_n R_i] = \sum_{n=\alpha}^{\beta} (^n\bar{K}_1[R])^i [M_n]/i! \qquad (9.23)$$

where the Poisson distribution of solubilizates is assumed for micelles of different aggregation number. Hence, the total solubilizate concentration becomes

$$[R_t] = [R] + \sum_{n=\alpha}^{\beta} {}^n\bar{K}_1[R][M_n] \exp({}^n\bar{K}_1[R]) \qquad (9.24)$$

and the following expression results:

$$([R_t] - [R])/[R] = \sum_{n=\alpha}^{\beta} {}^n\bar{K}_1[M_{n,t}] \qquad (9.25)$$

However, the difference between expressions (9.22) and (9.25) will disappear if the following operations for averaging are performed:

$$\bar{K}_1 = \sum_{n=\alpha}^{\beta} {}^n\bar{K}_1[M_{n,t}]/[M_t] \qquad (9.26)$$

$$[M_t] = \sum_{n=\alpha}^{\beta} [M_{n,t}] = (C - \text{CMC})/\bar{n} \qquad (9.27)$$

where $\bar{n}$ is the mean aggregation number. Thus, the $\bar{K}_1$ value is a mean value over the aggregation number of micelles.

As mentioned, the Poisson distribution shows excellent agreement with reality as long as the extent of solubilization is small. It should be stressed, however, that the distribution of solubilizates among micelles is determined by their association constants with micelles, not by mathematics. At solubilities to which the Poisson distribution is not applicable, the situation should be discussed from the viewpoint of microemulsion.

9.5. Factors Influencing Solubilization

It is very enlightening and useful to consider the factors influencing solubilization. The total solubilizate concentration $[R_t]$ is a function of $\bar{K}_1$, $[R]$, $[M]$ or $[M_t]$ from Eqs. (9.15) and (9.16). These values determine the general behavior of solubilization as follows:

1. The MAC or $[R_t]$ increases with increasing total surfactant concentration, as is clear from (9.16). As long as the value of $\bar{K}_1$ and [R] remain constant because solubilization is small, the MAC increases linearly above the CMC.[27-33]

2. The MAC of a hydrophobic solubilizate per mole of surfactant increases with alkyl chain length for a series of homologous surfactants.[23,32,34] This effect can be attributed to the increase in the number of surfactant molecules available for micellization caused by (1) the decrease in CMC; (2) an increase in $\bar{K}_1$ due to the increasing volume of the hydrophobic part of the micelles; and (3) the fortified hydrophobicity of the micelle palisade layer brought about by the closer packing of surfactant molecules at the micellar surface. The MAC also depends on the hydrophobicity of the solubilizate: the MAC decreases with an increase in the hydrophobicity or alkyl chain length of the solubilizates as a result of a decrease in [R]. However, $\bar{K}_1$ increases.[7,24,33] As can be seen from Table 9.1, the effect per methylene group of solubilizate is twice as great as that of the surfactant.[23]

3. The MAC of some solubilizates increases with temperature owing to an increase in the monomeric solubilizate concentration [R]. However, the $\bar{K}_1$ value ususally decreases with temperature.[19,20,35] In this case, the MAC depends largely on the temperature dependence of [R] and $\bar{K}_1$.

4. The MAC is increased by addition of excess salt.[12,36-38] This effect is due to (1) a decrease in CMC (as mentioned above), (2) an increase in micelle size and/or a change in micelle shape,[28] and (3) fortified hydrophobicity of the palisade layer owing to reduced repulsion between hydrophilic groups. These factors lead to the higher $\bar{K}_1$ value. However, the salt effect is relatively small for micelles with small aggregation numbers, for example bile salts.[39]

Table 9.1. Association Constant $(\bar{K}_1)$ and Average Number of Solubilizate Molecules per Micelle $(\bar{R})$ of Alkylsulfonic Acids[a]

	PTH		MPTH		EPTH	
Surfactant	$10^{-5}\,\bar{K}_1$, mol $\cdot$ dm^{-3}	$\bar{R}$	$10^{-5}\,\bar{K}_1$, mol $\cdot$ dm^{-3}	$\bar{R}$	$10^{-5}\,\bar{K}_1$, mol $\cdot$ dm^{-3}	$\bar{R}$
C_{12}-acid	0.664	0.70	1.44	0.65	2.21	0.34
C_{14}-acid	1.49	1.49	3.16	1.33	6.48	0.78
C_{16}-acid	2.29	2.30	6.78	2.28	11.74	1.27

[a] Reproduced with permission of Academic Press.

The MAC is determined in exactly the same way as solubility, by a measurement requiring complete separation of the micellar solution from the solubilizate phase. Other useful information on solubilization is available from liquid chromatography,[40] gel filtration,[41] NMR,[42-44] dialysis,[45] potentiometry,[46] spectroscopy,[47-51] mixed micelle formation,[52-54] kinetics,[55-58] and volumetric studies.[59,60]

9.6.· Location of Solubilizates in Micelles

- The position of solubilizates in micelles, as well as in living membranes, provides very important information concerning the physicochemical properties and physiological functions of both solubilizate and micelle or membrane. This property can be investigated using probe molecules, the molecular spectrum of which indicates the surrounding conditions.[61-65]

- The absorption spectrum of a molecule depends on the dielectric constant of the medium surrounding the molecule. The dielectric constant of a micelle ranges from 2 for the liquid hydrocarbon in the inner core to 80 for the water of the outer micellar surface. The following generally accepted rules for solubilizate position are derived from many works: (1) Nonpolar aliphatic hydrocarbons locate in an inner hydrophobic micellar core. (2) Semipolar and polar compounds such as alcohols, acids, and amines locate at the so-called polisade layer of the micelle with the polar group at the micellar surface and the nonpolar hydrocarbon groups in the micellar core. (3) Aromatic hydrocarbons such as benzene, toluene, and naphthalene sit in the micellar core and at the micellar surface (the two-state model).[62,66,67] The fraction of solubilizates occupying each of these two sites depends on the solubilizate concentration in the intermicellar bulk phase. The fraction in the inner core increases with increasing concentration in the bulk phase because the increase in the chemical potential of the solubilizate enables the solubilizate to move toward the micelle core.

Recently, the postulated Laplace pressure increase in the micellar interior has often been employed to explain the diminished free energy change of transfer per methylene group from the aqueous bulk into the micellar interior, compared with the free energy change from the aqueous bulk into bulk liquid hydrocarbon.[14,31,33,62,68] In addition, the decreased free energy change of micelle formation per methylene group has been attributed to partial crystallinity of the alkyl chain in the micellar interior caused by the increased pressure.[69] An interfacial tension does exist just at the boundary betwen two bulk phases.[70-72] If the postulated pressure increase exists, it can take place only in the case where micelles can be a separate phase. However, as mentioned in Chapter 4, according to the phase rule, micelles are not a separate phase but a chemical species. This model implies partial crystallinity of the alkyl chains in the solubilized state in the micellar core, with the hydrophilic group of solubilizate molecules anchored to the micellar surface, resulting in greatly reduced alkyl chain mobility and thus reduced free energy change of transfer per methylene group. In other words, the Laplace pressure is unnecessary to eluciate the above phenomena, which were investigated by the solubilization of p-n-alkylbenzoic acids with different alkyl chain lengths into dodecyl sulfonic acid micelles.[24]

References

1. M. E. L. McBain and E. Hutchinson, *Solubilization and Related Phenomena*, Academic Press, New York (1955).
2. P. H. Elworthy, A.T. Florence, and C. B. Macfarlane, *Solubilization by Surface-Active Agents and its Application in Chemistry and the Biological Sciences*, Chapman & Hall, London (1968).
3. R. L. Kroc, D. G. Whedon, and W. Garey, *The Physical Chemistry of Bile in Health and Disease* [*Hepatology*, 4(5)], Williams & Wilkins, Baltimore (1984).
4. J. W. Larsen and L. J. Magid, *J. Phys. Chem.* 78, 834 (1974).
5. K. Ogino, M. Abe, and N. Takeshita, *Bull. Chem. Soc. Jpn.* 49, 3679 (1976).
6. J. Gettins, D. Hall, P. L. Jobling, J. E. Rassing, and E. Wyn-Jones, *J. Chem. Soc. Faraday Trans. 2*, 74, 1957 (1978).
7. C. A. Bunton and L. Sepulveda, *J. Phys. Chem.* 83, 680 (1979).
8. E. Lissi, E. Abuin, and A. M. Rocha, *J. Phys. Chem.* 84, 2406 (1980).
9. C. H. Spink and S. Colgan, *J. Dolloid Interface Sci.* 97, 41 (1984).
10. S. M. Blokhus, H. Hoiland, and S. Backlund, *J. Colloid Interface Sci.* 114, 9 (1986).
11. A. E. Christian, E. E. Tucker, and E. H. Lane, *J. Colloid Interface Sci.* 84, 423 (1981).
12. E. D. Tucker and S. D. Christian, *J. Colloid Interface Sci.* 104, 562 (1985).
13. V. Rizzo, *J. Colloid Interface Sci.* 110, 110 (1986).
14. S. D. Christian, E. E. Tucker, G. A. Smith, and D. S. Bushong, *J. Colloid Interface Sci.* 113, 439 (1986).
15. E. Ruckenstein and R. Krishnan, *J. Colloid Interface Sci.* 71, 321 (1979).
16. R. Mallikarjun and D. B. Dadyburjor, *J. Colloid Interface Sci.* 84, 73 (1981).
17. K. S. Birdi and A. Ben-Naim, *J. Chem. Soc. Faraday Trans. 1*, 77, 741 (1981).
18. Y. Moroi, *J. Phys. Chem.* 84, 2186 (1980).
19. Y. Moroi, K. Sato, and R. Matuura, *J. Phys. Chem.* 86, 2463 (1982).
20. Y. Moroi, H. Noma, and R. Matuura, *J. Phys. Chem.* 87, 872 (1983).
21. Y. Moroi, K. Sato, H. Noma, and R. Matuura, in: *Surfactants in Solution* (K. L. Mittal and B. Lindman, eds.), Vol. 2, pp. 963–979, Plenum Press, New York (1984).
22. Y. Moroi, *J. Colloid Interface Sci.* 122, 308 (1988).
23. Y. Moroi and R. Matuura, *J. Colloid Interface Sci.* 125, 456 (1988).
24. Y. Moroi and R. Matuura, *J. Colloid Interface Sci.* 125, 463 (1988).
25. H. D. Young, *Statistical Treatment of Experimental Data*, McGraw-Hill, New York (1962).
26. W. Feller, *An Introduction of Probability Theory and Its Applications*, 3rd ed., Vol. 1, p. 35, Wiley, New York (1967).
27. H. Schott, *J. Phys. Chem.* 70, 2966 (1966).
28. S. Ozeki and S. Ikeda, *J. Phys. Chem.* 89, 5088 (1985).

29. T. Imae, A. Abe, Y. Taguchi, and S. Ikeda, *J. Colloid Interface Sci. 109*, 567 (1986).
30. P. T. Jacob and E. W. Anacker, *J. Colloid Interface Sci. 43*, 105 (1973).
31. I. B. C. Matheson and A. D. King, Jr., *J. Colloid Interface Sci. 66*, 464 (1978).
32. D. W. Ownby and A. D. King, Jr., *J. Colloid Interface Sci. 101*, 271 (1984).
33. W. Prapaitrakul nd A. D. King, Jr., *J. Colloid Interface Sci. 106*, 186 (1985).
34. M. Abu-Hamdiyyah and I. A. Rahman, *J. Phys. Chem. 91*, 1530 (1987).
35. K. S. Birdi, H.N. Singh, and S. U. Dalsager, *J. Phys. Chem. 83*, 2733 (1979).
36. J. C. Hoskins and A. D. King, Jr., *J. Colloid Interface Sci. 82*, 264 (1981).
37. Y. Nemoto and H. Funahashi, *J. Colloid Interface Sci. 80*, 542 (1981).
38. M. Fromon, A. K. Chattopadhyay, and C. Treiner, *J. Colloid Interface Sci. 102*, 14 (1984).
39. S. D. Christian, L. S. Smith, D. S. Bushoung, and E. E. Tucker, *J. Colloid Interface Sci. 89*, 514 (1982).
40. E. Pramauro, G. Saini, and E. Pelizzetti, *Anal. Chim. Acta 166*, 233 (1984).
41. A. Goto, M. Nihei, and F. Endo, *J. Phys. Chem. 84*, 2268 (1980).
42. P. Stilbs, *J. Colloid Interface Sci. 80*, 608 (1981).
43. R. E. Stark, R. W. Storrs, and M. L. Kasakevich, *J. Phys. Chem. 89*, 272 (1985).
44. S. Ghosh, M. Petrin, and A. H. Maki, *J. Phys. Chem. 90*, 5206 (1986).
45. G. A. Smith, S. D. Christian, E. E. Tucker, and J. F. Scamehorn, *J. Solution Chem. 15*, 519 (1986).
46. E. Azaz and M. Donbrow, *J. Phys. Chem. 81*, 1636 (1977).
47. H. Akasu, A. Nishi, M. Ueno, and K. Megro, *J. Colloid Interface Sci. 54*, 278 (1976).
48. J. B. S. Bonilha, T. K. Foreman, and D. G. Whitten, *J. Am. Chem. Soc. 104*, 4215 (1982).
49. E. B. Abuin and E. A. Lissi, *J. Colloid Interface Sci. 95*, 198 (1983).
50. J. C. Russell, U. P. Wild, and D. G. Whitten, *J. Phys. Chem. 90*, 1319 (1986).
51. L. B. Shih and R. W. Williams, *J. Phys. Chem. 90*, 1615 (1986).
52. F. Tokiwa, *J. Colloid Interface Sci. 28*, 145 (1968).
53. J. C. Hoskins and A. D. King, Jr., *J. Colloid Interface Sci. 82*, 260 (1981).
54. C. Treiner, J.-F. Bocquet, and C. Pommier, *J. Phys. Chem. 90*, 3052 (1986).
55. J. A. Shaeiwitz, A. F.-C. Chan, E. L. Cussler, and D. F. Evans, *J. Colloid Interface Sci. 84*, 47 (1981).
56. Y. Miyashita and S. Hayano, *J. Colloid Interface Sci. 86*, 344 (1982).
57. P. Lianos, M.-L. Viriot, and R. Zana, *J. Phys. Chem. 88*, 1098 (1984).
58. V. C. Reinsborough and J. F. Holzwarth, *Can. J. Chem. 64*, 955 (1986).
59. N. Funasaki, S. Hada, and S. Neya, *J. Phys. Chem. 88*, 1243 (1984).
60. M. Manabe, S. Kikuchi, S. Katayama, S. Tokunaga, and M. Koda, *Bull. Chem. Soc. Jpn. 57*, 2027 (1984).
61. P. Mukerjee, in: *Solution Chemistry of Surfactants* (K. L. Mittal, ed.), Vol. 1, pp. 153–174, Plenum Press, New York (1979).
62. P. Mukerjee and J. R. Cardinal, *J. Phys. Chem., 82*, 1620 (1978).
63. J. R. Cardinal and P. Mukerjee, *J. Phys. Chem. 82*, 1614 (1978).
64. P. Mukerjee, C. Ramachandran, and R. A. Pyter, *J. Phys. Chem. 86*, 3189 (1982).
65. C. Ramachandran, R. A. Pyter, and P. Mukerjee, *J. Phys. Chem. 86*, 3198, 3206 (1982).
66. K. Kasatani, M. Kawasaki, H. Sato, and N. Nakashima, *J. Phys. Chem. 89*, 542 (1985).
67. L. Sepulveda, *J. Colloid Interface Sci. 46*, 372 (1974).
68. S. A. Simon, R. V. McDaniel, and T. J. McIntosh, *J. Phys. Chem. 86*, 1449 (1982).
69. P. Mukerjee, *Kolloid Z. Z. Polym. 236*, 76 (1970).
70. J. L. Moilliet and B. Collie, *Surface Activity*, Van Nostrand, Princeton, N. J. (1950).
71. R. Defay, I. Prigogine, A. Bellemans, and D. H. Everret, *Surface Tension and Adsorption*, Longmans, London (1966).
72. E. A. Guggenheim, *Thermodynamics*, Elsevier/North-Holland, Amsterdam (1977).

10

Mixed Micelle Formation

10.1. Introduction

The incorporation of solubilizates into a surfactant micelle results in the formation of a *mixed micelle*. Solubilization is thus closely related to mixed micelle formation. As ususally used, however, the *mixed micelle* means a micelle composed of surfactants capable themselves of forming micelles. By this usage, mixed micellization is a special case of solubilization.

Many papers have appeared on mixed micelle formation[1-30] but almost all of them are based on the phase-separation rather than the mass-action model of micelles.[1,22] This is also true for solubilization. As has been emphasized repeatedly in the foregoing chapters, micelles are not a separate phase but a chemical species. Therefore, mixed micelles also should be treated as a chemical species. Unfortunately, the interpretations of mixed micellization based on the mass-action model have not agreed well with experimental CMC values,[1] probably because the physiocochemical properties of mixed micelles are quite different from those of pure micelles of the individual components. In addition, the micellar aggregation number and the association of counterions with micelles change dramatically with composition in mixed micelles, even though mixed micelles of homologous surfactants differing only in hydrophobic chain length are expected to have surface properties similar to those of pure micelles of each surfactant. In fact, theoretical mixed CMC values of homologous surfactants agree well with experimental values over the whole composition range.[2,4,12]

Even though micelles should be treated as a chemical species, a pseudophase model of the micelle is also of considerable practical importance for estimating CMC values of mixed surfactants. Therefore, the following discussion may be regarded as a practical effort to derive an empirical equation for mixed micelles. First, however, it is instructive to know how closely micelles resemble a separate phase, because almost all approaches to mixed micelle formation have depended on this approximation because the model discussed in the second section of this chapter holds. When the approximation is correct, the theoretical discussion of mixed micelles becomes analogous to the two-phase equilibrium; i.e., either a chemical equiibrium of surfactant monomers is estblished between the micellar phase and intermicellar bulk phase, or Raoult's law is obeyed when the latter phase is assumed to be a gaseous phase. This is expressed as

$$C_i = x_i C_i^0 \qquad (10.1)$$

where C_i is the concentration of component i at the mixed CMC, x_i the mole fraction of i in the mixed micelle, and C_i^0 the CMC of pure component i.

Figure 10.1 shows the change of the mixed CMC of dodecyl sulfonic acid with the composition of p-n-alkylbenzoic acid solubilizates.[31] If mixed micellization strictly followed Raoult's law, the CMCs of this series should lie on the dashed line in Fig. 10.1, but in fact almost all of the CMCs lie

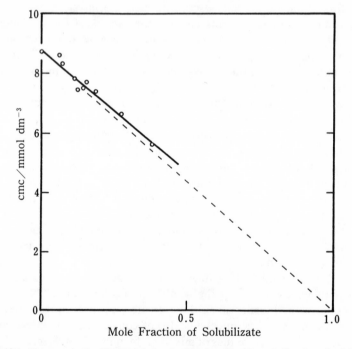

Figure 10.1. Decrease in the CMC of dodecyl sulfonic acid with mole fraction of solubilizates in the micelle.[31] (Reproduced with permission of Academic Press.)

slightly above it. Raoult's law is thus not strictly applicable. However, it provides a tolerable approximation as long as there is no specific interaction between solubilizate and surfactant (as is true in the present case).

10.2. Two-Component Surfactant Systems

The solution properties of a binary surfactant mixture fall either between or outside the solution properties of the two single-surfactant solutions. This is also true for the CMC of a binary surfactant solution. Most CMCs of binary surfactant mixtures fall between the CMCs of the two components, but some fall above[32-35] or below[3,12,16,17,29] this range.

Many of the theories concerning the CMC of binary surfactant mixtures have assumed the ideality of each component in the micellar phase.[2,4,15] These theories work well for binary mixtures of homologous surfactants but not for nonhomologous mixtures. Mixed micelles in solutions of non-homologous surfactant mixtures should be nonideal because the interaction between identical surfactants is different from that between nonhomologous ones. A simple way to take this nonideality into account is to employ regular solution theory. An approach using this theory can elucidate binary mixtures of nonhomologous surfactants quite well by using the single adjustable parameter developed by Rubingh.[16]

The chemical potential μ_1 of monomeric surfactant 1 in an intermicellar bulk phase is expressed in the ususal way as

$$\mu_1 = \mu_1^{\ominus} + RT \ln C_1 \tag{10.2}$$

where $\mu_1^{\ominus}$ is its standard chemical potential and C_1 is the concentration. On the other hand, the chemical potential of surfactant 1 in a micellar phase is written likewise as

$$\mu_1^m = \mu_1^{m,0} + RT \ln \gamma_1 x \tag{10.3}$$

where γ_1 and x is the activity coefficient due to nonideal mixing and x is the mole fraction of component 1 in the binary surfactant phase. When a micelle is made of a single component 1, the following relation results from Eqs. (10.2) and (10.3):

$$\mu_1^{m,0} = \mu_1^{\ominus} + RT \ln C_1^0 \tag{10.4}$$

where C_1^0 is the CMC of the single component 1. From the equilibrium of component 1 between the micellar phase and the intermicellar bulk phase, $\mu_1 = \mu_1^m$, we have

$$C_1 = \gamma_1 x C_1^0 \tag{10.5}$$

For component 2, we have a similar equation

$$C_2 = \gamma_2 (1 - x) C_2^0 \qquad (10.6)$$

At the CMC (C), the following relations are satisfied from the mass balances for components 1 and 2:

$$\alpha C = C_1 \quad \text{and} \quad (1 - \alpha)C = C_2 \qquad (10.7)$$

where α is the net mole fraction of component 1. Then, we have

$$\alpha C = \gamma_1 x C_1^0 \qquad (10.8)$$

$$(1 - \alpha)C = \gamma_2 (1 - x) C_2^0 \qquad (10.9)$$

Elimination of x from (10.8) and (10.9) leads to

$$1/C = \alpha / \gamma_1 C_1^0 + (1 - \alpha)/\gamma_2 C_2^0 \qquad (10.10)$$

where the case $\gamma_1 = \gamma_2 = 1$ represents ideal mixing.

On the other hand, elimination of C_t generates the relationship between x and α at the CMC:

$$x = \alpha \gamma_2 C_2^0 / [(1 - \alpha)\gamma_1 C_1^0 + \alpha \gamma_2 C_2^0] \qquad (10.11)$$

The relationship between x and total surfactant C above the CMC is derived by the mass balance. The value of x is easily given by

$$x = (\alpha C_t - C_1)/(C_t - C_1 - C_2) \qquad (10.12)$$

and substituting C_1 and C_2 from (10.5) and (10.6) into (10.12) results in

$$x = \{-(C_t - \Delta) + [(C_t - \Delta)^2 + 4\alpha C_t \Delta]^{1/2}\}/2\Delta \qquad (10.13)$$

where

$$\Delta = \gamma_2 C_2^0 - \gamma_1 C_1^0 \qquad (10.14)$$

From (10.5), (10.6), and (10.13) we obtain the monomer concentrations

$$C_1 = \{-(C_t - \Delta) + [(C_t - \Delta)^2 + 4\alpha C_t \Delta]^{1/2}\}/2[(\gamma_2 C_2^0 / \gamma_1 C_1^0) - 1] \qquad (10.15)$$

$$C_2 = \{-(C_t + \Delta) + [(C_t - \Delta)^2 + 4\alpha C_t \Delta]^{1/2}\}/2[(\gamma_1 C_1^0 / \gamma_2 C_2^0) - 1] \qquad (10.16)$$

If the values of C_1^0, C_2^0, γ_1, and γ_2 are available, x, C_1, and C_2 may be determined by calculation. The values of C_1^0 and C_2^0 are obtained from the CMC values of the single-surfactant solutions, but the activity coefficients are given by the regular solution theory as

$$\gamma_1 = \exp[\beta(1-x)^2] \tag{10.17}$$

$$\gamma_2 = \exp(\beta x^2) \tag{10.18}$$

where $\beta(=NZ\omega/RT)$ is the interaction parameter (Chapter 3). From the logarithms of (10.17) and (10.18), there results

$$\ln \gamma_1/\ln \gamma_2 = (1-x)^2/x^2 \tag{10.19}$$

Introducing γ_1 of (10.8) and γ_2 of (10.9) into (10.19) results in

$$x^2 \ln (\alpha C/xC_1^0)/\{(1-x)^2 \ln [(1-\alpha)C/(1-x)C_2^0]\} = 1 \tag{10.20}$$

At the same time, the β value can be evaluated by substituting (10.8) into (10.17):

$$\beta = \ln(\alpha C/xC_1^0)/(1-x)^2 \tag{10.21}$$

If the CMC of a binary mixture (C) is determined against the net mole fraction α, then the micellar composition x is given by (10.20), and β may be obtained from (10.21) by using the x value. In general, a single parameter value of β is determined by averaging the β values against each α value over the entire composition range. Figure 10.2 shows the change with composition of a binary surfactant solution, plotted using a single parameter value of β. Table 10.1 gives these β values for many binary surfactant mixtures. The β value is an index of interaction between two surfactants. A large negative value of β indicates strong interaction between the surfactants, whereas a positive value indicates repulsion. The sign of the β value corresponds to positive or negative deviation from ideality. Thus, the β values in Table 10.1 indicate strong interaction for binary mixtures of cationic with anionic surfactants, more negative values of β, moderate interaction between anionic and nonionic surfactants, and weak interaction between cationic and nonionic surfactants.

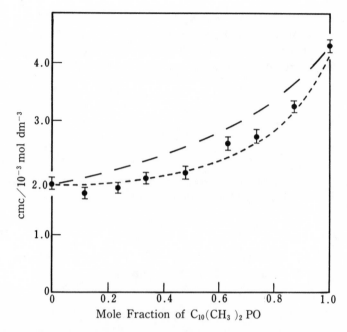

Figure 10.2. Mixed CMC values of $C_{10}H_{21}(CH_3)_2PO$ and $C_{10}H_{21}(CH_3)SO$ against the mole fraction of $C_{10}H_{21}(CH_3)_2PO$. ●, experimental data; – – –, ideal mixed micelle; - - - -, nonideal mixed micelle ($\beta = -0.84$).[16]

Table 10.1. β Values for Mixed Micelle Formation of Binary Systems[16]

System	β value
$C_{12}OSO_3^-Na^+/C_8E_6$	−4.1
$C_{12}OSO_3^-Na^+/C_{10}E_6$	−3.9
$C_{15}OSO_3^-Na^+/C_{10}E_6$	−4.3
C_{12}—⟨○⟩—$SO_3^-Na^+/C_{12}NO(CH_3)_2$	−3.5
$C_{16}N^+(CH_3)_3Cl^-/C_{12}E_5$	−2.4
$C_{14}N^+(CH_3)_3Cl^-/C_{10}E_5$	−1.5
$C_{14}N^+(CH_3)_2CH_2$—⟨○⟩—$Cl^-/C_{10}E_5$	−1.5
$C_{20}N^+(CH_3)_3Cl^-/C_{12}E_8$	−4.6
$C_{10}(CH_3)_2PO/C_{10}(CH_3)SO$	−0.84
$C_{12}N^+(CH_3)_3Cl^-/C_{12}N^+(CH_3)_2(CH_2)_3SO_3^-$	−1.0
$C_{12}OSO_3^-Na^+/C_{12}N^+(CH_3)_2(CH_2)_3SO_3^-$	−7.8
$C_{10}OSO_3^-Na^+/C_{10}N^+(CH_3)_3Br^-$	−18.5
$C_{12}(CH_3)_2PO/C_{12}N^+(CH_3)_2(CH_2)_5COO^-$	−1.0

10.3. Partially Miscible Micelles and Demicellization

The β values in Table 10.1 are all negative, indicating that the interaction between these surfactant pairs is larger than the interaction in the pure compound. This section considers what takes place when the β value is positive.[32,36] As is well known, phase separation occurs in a binary solution when the interaction of the binary components exceeds the critical value $\beta = 2$ for the regular solution theory.[37] Similarly, a binary surfactant solution contains two kinds of micelles of different composition.[32-35,38-41] For example, mixed solutions of hydrocarbon and fluorocarbon (typical nonpolar hydrophobic substances) are far from ideality,[42] strongly suggesting that these binary surfactant solutions contain two kinds of mixed micelles: one rich in hydrocarbon and the other rich in fluorocarbon.

Figure 10.3 shows the CMC values of mixtures of sodium perfluorooctanoate (SPFO) with sodium dodecanoate (SD) and with sodium decyl sulfate (SDeS) in a 0.001 N NaOH aqueous solution. Both mixtures show two CMCs, and the first CMC is higher than the value expected from ideal mixing of the two kinds of chain moieties. The first CMC represents the point at which the surfactant present in higher concentration begins to micellize, and the second CMC represents the concentration at which the

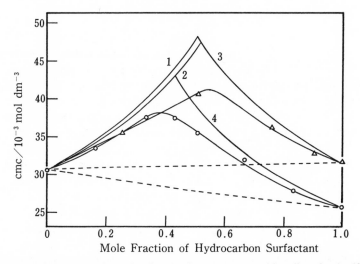

Mole Fraction of Hydrocarbon Surfactant

Figure 10.3. Mixed CMC values of sodium perfluorooctanoate with sodium decyl sulfate ($\triangle$) and sodium dodecanoate ($\bigcirc$).[32] – – – –, ideal mixed micelle with $B = 0.645$ [Eq. (10.23)]; 2–4, complete demixing of micelles with $B = 0.645$ [Eq. (10.25)]; 1, complete demixing of micelles with $B = 0.53$ [Eq. (10.25)]. (Reproduced with permission of the American Chemical Society.)

second surfactant begins to micellize. Thus, two kinds of micelles coexist above the second CMC. When the two micelles are completely demixed, the first CMC can be calculated by the condition that the concentration of either surfactant attains its CMC [Eq. (10.7)]:

$$\alpha C = C_1^0 \tag{10.22}$$

On the other hand, owing to the effect of counterion concentration, the CMC of ionic surfactants changes in accordance with the following equation:

$$\log \alpha C = \text{const} - B \log C \tag{10.23}$$

where β is the degree of counterion association to the micelle. This relation is also the case for a single surfactant

$$\log C_1^0 = \text{const} - B \log C_1^0 \tag{10.24}$$

From (10.23) and (10.24), there results

$$\log(C/C_1^0) = \log(1/\alpha)/(1 + B) \tag{10.25}$$

The theoretical first CMC values at reasonable values of β are given by the solid lines in Fig. 10.3. Although the theoretical values do not agree well with the experimental values, they clearly indicate the severe nonideality of mixing and the presence of two kinds of micelles. Shinoda and Nomura explained the positive deviation by applying the regular solution theory,[38] but the explanation, which neglected the presence of the second CMC, was not reasonable. Mysels further developed the concept of coexistence of two kinds of micelles postulated by Mukerjee and Yang, and introduced the critical demicellization concentration.[36] When the total surfactant concentration is gradually increased in a binary surfactant solution containing two kinds of micelles (keeping the surfactant composition constant), a certain composition range is found at which one kind of micelle disappears because it is solubilized into the other micelles, which increase in number with total surfactant concentration. The total surfactant concentration at which one kind of micelle disappears completely is called the *critical demicellization concentration*. This phenomenon has been demonstrated by Funasaki and Hada.[41]

10.4. Multicomponent Surfactant Systems

Surfactants used in practical applications are ususally multicomponent rather than binary mixtures. Various papers treating multicomponent surfactant mixtures in terms of mixed micelle formation have appeared.[5,12,17,20] In this section, the topic is treated from the same standpoint as in Section 10.2.[17]

For a multicomponent surfactant mixture, Eq. (10.8) can be written as

$$C_i = \alpha_i C = \gamma_i x_i C_i^0 \qquad (10.26)$$

where α_i is the net mole fraction of component i, and γ_i, x_i, and C_i are the variables corresponding to those in Section 10.2, at the mixed CMC C. From the constraint $\sum_i x_i = 1$, Eq. (10.26) gives

$$1/C = \sum_i^n \alpha_i/(\gamma_i C_i^0) \qquad (10.27)$$

for multicomponent micelles of n total surfactants. Mass balance of component i leads to a relationship between the net mole fraction α_i and the mole fraction in the mixed micelle x_i

$$x_i = (\alpha_i C_t - C_i)/(C_t - C) \qquad (10.28)$$

Substitution of (10.26) into (10.28) results in

$$x_i = \alpha_i C_t/(C_t + \gamma_i C_i^0 - C) \qquad (10.29)$$

for the mole fraction x_i. Substitution of (10.29) into (10.26) then gives the monomeric concentration of component i:

$$C_i = \alpha_i \gamma_i C_i^0 C_t/(C_t + \gamma_i C_i^0 - C) \qquad (10.30)$$

From Eqs. (10.29) and (10.30), x_i and C_i can be obtained if γ_i is known because the other parameters are all experimentally determined.

The next question is how to evaluate the γ_i values. Two equations of (10.26) with respect to components i and j yield the following equation at the CMC:

$$x_i = \alpha_i C_j^0 \gamma_j x_j/C_i^0 \alpha_j \gamma_i \qquad (10.31)$$

At the same time, (10.29) is also applicable to component x_j

$$x_j = \alpha_j C_t/(C_t + \gamma_j C_j^0 - C) \qquad (10.29')$$

Elimination of C from (10.29) and (10.29') results in

$$x_i = \alpha_i C_t/(\gamma_i C_i^0 - \gamma_j C_j^0 + \alpha_j C_t/x_j) \qquad (10.32)$$

These expressions provide basic relations to solve the activity coefficients γ_i. For binary surfactant mixtures, the activity coefficient can be easily expressed in terms of one mole fraction, as is seen in Section 10.2. For multicomponent mixtures, the activity coefficient is more complex:

$$\ln \gamma_i = \sum_{\substack{j=1 \\ (j \neq i)}}^{n} \beta_{ij} x_j^2 + \sum_{\substack{j=1 \\ (i \neq j \neq k)}}^{n} \sum_{k=1}^{j-1} (\beta_{ij} + \beta_{ik} - \beta_{jk}) x_j x_k \qquad (10.33)$$

where β_{ij} represents the pairwise interaction between components i and j. The number of pairwise interactions is given by the combination of C_2^n. Equation (10.33) also contains an equivalent number of terms of the activity coefficient γ_i, with $n - 1$ direct terms containing $x_j^2 (i = j)$ and $C_2^n - n + 1$ cross terms containing $x_i x_j (i \neq j \neq k)$. The pairwise interaction parameters β_{ij} can be independently determined from binary mixtures.

The next problem for nonideal multicomponent mixtures is to solve the n activity coefficients for the x_i values at the total surfactant composition and concentration. To solve the n activity coefficients and the n mole fractions, we need $2n$ equations: n equations of (10.33) and n equations of (10.31) or (10.32), with the constraint that the sum of the x_i values equals unity. A numerical solution of multiple equations for multiple unknowns can be reached efficiently using the Nelder–Mead simplex technique.[43] Once the γ_i values have been determined, the mole fraction in micelle x_i and the monomer concentration C_i for a multicomponent surfactant solution are easily determined by (10.29) and (10.30). The former values are, of course, obtained together with the γ_j values. Figure 10.4 shows the CMCs determined by this procedure for the ternary mixture of $C_{10}H_{21}(CH_3)_2PO/C_{10}H_{21}(CH_3)SO/C_{12}H_{25}SO_4Na$. For this ternary mixture, Eq. (10.33) is written as

$$\ln \gamma_1 = \beta_{12} x_2^2 + \beta_{13} x_3^2 + (\beta_{12} - \beta_{23} + \beta_{13}) x_2 x_3 \qquad (10.34)$$

$$\ln \gamma_2 = \beta_{23} x_3^2 + \beta_{12} x_1^2 + (\beta_{23} - \beta_{13} + \beta_{12}) x_3 x_1 \qquad (10.34')$$

$$\ln \gamma_3 = \beta_{13} x_1^2 + \beta_{23} x_2^2 + (\beta_{13} - \beta_{12} + \beta_{23}) x_1 x_2 \qquad (10.34'')$$

Mixed micellar solutions exhibit some very interesting properties not expected from individual surfactant solutions. The degree of counterion association to an ionic micelle is about 0.7 for monovalent counterions and 0.9 for divalent counterions. When an ionic surfactant is mixed with a nonionic surfactant, the degree of the association falls to zero as the mole fraction of nonionic surfactant in the micelle increases.[13,44-47] This is particularly evident for mixtures of anionic and nonionic surfactants of the

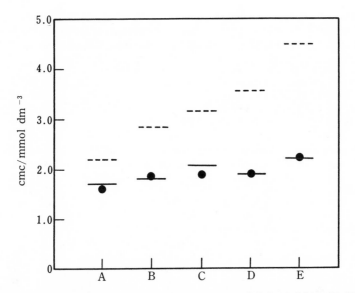

Figure 10.4. Mixed CMC of the ternary system of $C_{10}H_{21}(CH_3)_2PO$ (1)/$C_{10}H_{21}(CH_3)SO$ (2)/$C_{12}H_{25}SO_4Na$ (3) in 1 mM Na_2CO_3 at 24°C.[17] Molar ratios: A, 0.201/0.646/0/153; B, 0.354/0.378/0.268; C, 0.622/0.221/0.157;; D, 0.231/0.246/0.523; E, 0.136/0.145/0.719. ●, experimental data; - - - -, ideal mixed micelle; ———, nonideal mixed micelle ($\beta_{12} = 0$, $\beta_{13} = -3.7$, $\beta_{23} = -2.4$). (Reproduced with permission of the American Chemical Society.)

polyoxyethylene type, because of the strong interaction between the anionic head group and the ethylene oxice group.[48,49]

Various papers have also been published that treat surfactant mixtures from the viewpoints of foam stability,[50] gel filtration,[51] surface adsorption on fibers,[52] nuclear magnetic resonance,[53] light scattering,[54,55] and excess entropy.[46]

References

1. H. Lange, *Kolloid Z. 131*, 96 (1953).
2. H. Lange and K.-H. Beck, *Kolloid Z. Z. Polym. 251*, 424 (1973).
3. C.-P. Kurzendorfer, M.-J. Schwuger, and H. Lange, *Ber. Bunsenges. Phys. Chem. 82*, 962 (1978).
4. K. Shinoda, *J. Phys. Chem. 58*, 541 (1954).
5. K. Shinoda, *J. Phys. Chem. 58*, 1136 (1954).
6. K. J. Mysels and R. J. Otter, *J. Colloid Sci. 16*, 474 (1961).
7. L. Shedlovsky, C. W. Jakob, and M. B. Epstein, *J. Phys. Chem. 67*, 2075 (1963).
8. M. J. Schick and D. J. Manning, *J. Am. Oil Chem. Soc. 43*, 133 (1966).
9. M. J. Schick, *J. Am. Oil Chem. Soc. 43*, 681 (1966).
10. F. Tokiwa and N. Moriyama, *J. Colloid Interface Sci. 30*, 338 (1969).

11. Y. Moroi, K. Motomura, and R. Matuura, *J. Colloid Interface Sci.* 46, 111 (1974).
12. Y. Moroi, N. Nishikido, and R. Matuura, *J. Colloid Interface Sci.* 50, 344 (1975).
13. Y. Moroi, N. Nishikido, M. Saito, and R. Matuura, *J. Colloid Interface Sci.* 52, 356 (1975).
14. H. Maeda, M. Tsunoda, and S. Ikeda, *J. Phys. Chem.* 78, 1086 (1974).
15. J. H. Clint, *J. Chem. Soc. Faraday Trans. 1,* 76, 1327 (1975).
16. D. N. Rubingh, in: *Solution Chemistry of Surfactants* (K. L. Mittal, ed.), Vol. 1, p. 337, Plenum Press, New York (1979).
17. P. M. Holland and D. N. Rubingh, *J. Phys. Chem.* 87, 1984 (1983).
18. N. Funasaki, *J. Colloid Interface Sci.* 67, 384 (1978).
19. N. Funasaki and S. Hada, *J. Phys. Chem.* 83, 2471 (1979).
20. F. Harusawa and M. Tanaka, *J. Phys. Chem.* 85, 882 (1981).
21. R. F. Kamrath and E. I. Franses, *Ind. Eng. Chem. Fundam.* 22, 230 (1983).
22. R. F. Kamrath and E. I. Franses, *J. Phys. Chem.* 88, 1642 (1984).
23. K. Motomura, M. Yamanaka, and M. Aratono, *Colloid Polym. Sci.* 262, 948 (1984).
24. M.J. Hey, J.W. MacTaggart, and C.H. Rochester, *J. Chem. Soc. Faraday Trans. 1,* 81, 207 (1985).
25. R. Nagarajan, *Langmuir 1,* 331 (1985).
26. D. G. Hall and R. W. Huddleston, *Colloids Surfaces 13,* 209 (1985).
27. P. M. Holland, *Adv. Colloid Interface Sci.* 26, 111 (1986).
28. P. M. Holland, *Colloids Surfaces 19,* 171 (1986).
29. C. M. Nguyen, J. F. Rathman, and J. F. Scamehorn, *J. Colloid Interface Sci.* 112, 438 (1986).
30. J. F. Rathman and J. F. Scamehorn, *Langmuir 2,* 354 (1986).
31. Y. Moroi and R. Matuura, *J. Colloid Interface Sci.* 125, 463 (1988).
32. P. Mukerjee and A. Y. S. Yang, *J. Phys. Chem.* 80, 1388 (1976).
33. N. Funasaki and S. Hada, *J. Phys. Chem.* 84, 736 (1980).
34. T. Asakawa, K. Johten, S. Miyagishi, and M. Nishida, *Langmuir 1,* 347 (1985).
35. G. Sugihara, M. Yamamoto, Y. Wada, Y. Murata, and Y. Ikawa, *J. Solution Chem.* 17, 225 (1988).
36. K. J. Mysels, *J. Colloid Interface Sci.* 66, 331 (1978).
37. E. A. Guggenheim, *Mixture,* Oxford University Press (Clarendon), London (1952).
38. K. Shinoda and T. Nomura, *J. Phys. Chem.* 84, 365 (1980).
39. N. Funasaki and S. Hada, *J. Phys. Chem.* 86, 2504 (1982).
40. M. Abe, N. Tsubaki, and K. Ogino, *J. Colloid Interface Sci.* 107, 503 (1985).
41. N. Funasaki and S. Hada, *J. Colloid Interface Sci.* 73, 425 (1980).
42. J. H. Hildebrand, J. M. Prausnit, and R. L. Scott, *Regula and Related Solutions,* Chapter 10, Van Nostrand-Reinhold, Princeton, N.J. (1970).
43. D. M. Olsson, *J. Qual. Technol.* 6, 53 (1974).
44. M. Meyer and L. Sepulveda, *J. Colloid Interface Sci.* 99, 536 (1984).
45. J. F. Rathman and J. F. Scamehorn, *J. Phys. Chem.* 88, 5807 (1984).
46. I. W. O. Lee, R. S. Schechter, W. H. Wade, and Y. Barakat, *J. Colloid Interface Sci.* 108, 60 (1985).
47. M. Jansson and R. Rymden, *J. Colloid Interface Sci.* 119, 185 (1987).
48. Y. Moroi, H. Akisada, M. Saito, and R. Matuura, *J. Colloid Interface Sci.* 61, 233 (1977).
49. N. Nishikido, *J. Colloid Interface Sci.* 60, 242 (1977).
50. M. J. Schick and F. M. Fowkes, *J. Phys. Chem.* 61, 1062 (1957).
51. F. Tokiwz, K. Ohki, and I. Kokubo, *Bull. Chem. Soc. Jpn.* 41, 2845 (1968).
52. Y. Iwadare, *Bull. Chem. Soc. Jpn.* 43, 3364 (1970).
53. F. Tokiwa and K. Tsujii, *J. Phys. Chem.* 75, 3560 (1971).
54. P. Guering, P.-G. Nilsson, and B. Lindman, *J. Colloid Interface Sci.* 105, 41 (1985).
55. M. Corti, V. Degiorgio, R. Ghidoni, and S. Sonnino, *J. Phys. Chem.* 86, 2533 (1982).

11

Micellar Catalysis

11.1. Effects of Micelles on Chemical Reactions

Most important reactions occur not in a homogeneous solution but at an interface. Many industrially important processes occur on the surfaces of solid catalysts, and nearly all biological reactions take place at gas–liquid interfaces or on an enzyme that may itself be bound to a membrane. The properties of these catalytic surfaces depend critically on the detailed structure of the surface, which can be controlled by adding agents that may themselves take no direct part in the chemical reactions.[1]

These comments are applicable to micelle-catalyzed reactions. A solution containing micellar aggregates is macroscopically homogeneous (i.e., is one phase) by the ususal criteria of physical chemistry, as discussed in previous chapters. However, microscopically this phase is separated into many small regions of high solute concentration (micelles) dispersed in a solvent region. Any reactive species added to the solution will distribute itself between these regions. If the conditions in these two environments result in different reaction rates, then the micelles will act as either catalysts or inhibitors.[2] The catalytic efficiency will be governed both by the affinity of the reagents for the micelles and by the reactivity of the bound reagent molecules.

Micelles in aqueous media have either a polar region or a region of high charge density, accompanied by an electrostatic potential of up to a few hundred millivolts at the micellar surface, and a nonpolar hydrophobic region in the micelle core. Micelle aggregation numbers ususally range from less than 100 for ionic surfactants to several hundred for nonionic surfactants. Therefore, the kinetics of micellar solutions is governed by electrostatic and hydrophobic interactions between micelles and reactants, transition complexes, and products. If any of the reaction species interacts with micelles, then the presence of micelles will affect the reaction rate.

Various kinetic studies of micellar catalysis have examined the following types of micellar catalysis: (1) reactions in which the micelles are reagents; (2) reactions in which interactions between the micelles and the reacting species affect the kinetics; and (3) reactions in which the micelles carry catalytically active substituents.[2] These studies have been undertaken to elucidate the factors that influence the rates and courses of reactions, to gain insight into the exceptional catalytic characteristics of enzymatic reactions, and to explore the usefulness of micellar systems for organic synthesis.[3]

Micelle-catalyzed reactions are somewhat similar to enzyme-catalyzed reactions; the proper choice of surfactant brings about a rate increase of up to 1000-fold, and the diameter of micelles is 30–50 Å, similar in size to globular enzymes. Micelle-catalyzed reactions can be treated in a manner analogous to the reaction scheme for enzymatic catalysis, as will be shown below.

However, the analogy between the high reactivity of an enzyme-substrate complex and the reactivity of substrates bound to micelles is not entirely satisfactory. Extensive study of enzymatic reactions has demonstrated that the affinity of enzymes for small substrate molecules depends critically on the spacing of the interacting groups in the small molecules to be bound.[4] In a substrate–enzyme complex the reactions are fixed in position, whereas reactants incorporated into a micelle are free to move about in the micellar region. Furthermore, reactants distribute into micelles according to their solubilities and not according to the stoichiometry of the reaction. The solubilization of reactants and their distribution among micelles play the most important role in micelle-catalyzed reactions, as should be obvious for the case where the reactant concentrations are greater than the micellar concentration at surfactant concentrations near the CMC. In addition, the rate enhancement generally increases with increasing hydrophobicity of reactants and amphiphiles, which is not always the case for enzymatic reactions. These differences are mainly due to the fact that micelles do not maintain a definite configuration but are in dynamic association–dissociation equilibrium with monomeric surfactants in the bulk phase, changing their size and shape at rates of milliseconds and microseconds (see Section 4.7).

The electrostatic surface potential at the micellar surface can attract or repel ionic reaction species, and a strong hydrophobic interaction can bring about the incorporation into micelles even of reagents that bear the same charge as ionic micelles.[5] The number of reagent molecules per micelle can often be controlled by adjusting the surfactant concentrations; and thus a chemical reaction can be induced to yield specific products by selecting the proper combination of reactants and surfactants. In fact, the research area of photochemistry in micellar systems is now rapidly expanding, as is discussed in Chapter 12.

The incorporation of reagents into micelles often alters the CMC of the surfactant. Therefore, the CMC must be determined for each reaction system in order to correctly interpret the results.

Over the past 20 years, many reports, review articles, and monographs have appeared concerning reaction kinetics in aqueous and nonaqueous solutions of ionic and nonionic surfactants.[1-3,6-10] The details of this topic exceed the scope of this chapter, and interested readers should refer to the literature.

11.2. Characteristics of Enzymatic Reactions

During the late 19th century, enzyme-catalyzed reactions were widely studied from the practical standpoint of fermentation. Invertase, the enzyme that hydrolyzes sucrose, was intensively investigated and proved to be a true catalyst. Enzymatic catalysis is not only highly efficient but also exhibits remarkable specificity. Enzymes are proteins composed of polypeptide chains with a well-defined conformation. The enzyme molecule is flexible to some extent, and it undergoes deformation during association with a substrate. In general, the high reactivity of the enzyme–substrate complex can be accounted for by the proximity of one or more fundamental groups of the enzyme to the sensitive bond of the substate. Investigation established the existence of the enzyme–substrate complex, and Michaelis and Menten proposed the following reaction mechanism, based on their study of invertase-catalyzed reaction by mutarotation[11]:

$$E + S \underset{k_2}{\overset{k_1}{\rightleftharpoons}} ES \overset{k_3}{\longrightarrow} E + P \qquad (11.1)$$

where E, S, ES, and P refer respectively to enzyme, substrate, enzyme–substrate complex, and product, and k_1, k_2, and k_3 are the corresponding rate constants. The reversible first step is sufficiently rapid to be represented by an equilibrium constant (K_s)

$$K_s = k_2/k_1 = [E][S]/[ES] \qquad (11.2)$$

where the brackets denote the concentration. In terms of initial concentration of enzyme $[E_0]$, $[ES]$ is expressed as

$$[ES] = [E_0][S]/(K_s + [S]) \qquad (11.3)$$

The second reaction step is a simple first-order reaction with rate v:

$$v = k_3[\text{ES}] = k_3[\text{E}_0][\text{S}]/(K_s + [\text{S}]) \tag{11.4}$$

Together with Eq. (11.7), Eq. (11.4) is known as the *Michaelis–Menten equation*. Later, Briggs and Haldane derived a more generalized equation from the same standpoint[12]:

$$\text{d}[\text{ES}]/\text{d}t = k_1([\text{E}_0] - [\text{ES}])[\text{S}] - (k_2 + k_3)[\text{ES}] \tag{11.5}$$

   If the concentration of the complex is assumed to be steady state, we have

$$[\text{ES}] = k_1[\text{E}_0][\text{S}]/(k_2 + k_3 + k_1[\text{S}]) \tag{11.6}$$

Then,

$$v = V[\text{S}]/(K_m + [\text{S}]) \tag{11.7}$$

where K_m is the *Michaelis constant*, defined as $(k_2 + k_3)/k_1$, and V is the maximum velocity, defined as $k_3[\text{E}_0]$, in the presence of excess substrate. However, V is not a fundamental property of an enzyme, and the catalytic constant k_3 is preferable to V. Figure 11.1 shows a graphic form of (11.7).[13]

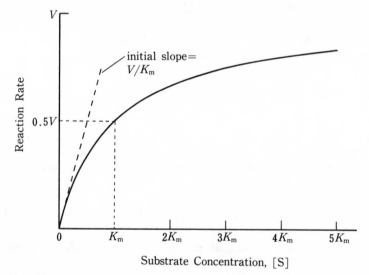

Substrate Concentration, [S]

Figure 11.1. Change of initial velocity v with substrate concentration [S] for a reaction obeying the Michaelis–Menten equation. (Reproduced with permission of Butterworths.)

The curve is a rectangular hyperbola with two asymptotes $[S] = -K_m$ and $v = V$. At very small values of $[S]$, v is directly proportional to $[S]$ ($v = V[S]/K_m$); thus, the initial slope is V/K_m. When $[S]$ is equal to K_m, the reaction rate is just half-maximal, i.e., $v = 0.5V$ at $[S] = K_m$. Thus, the reaction appears first order in $[S]$ at small values of $[S]$, but zero order at larger values of $[S]$. When the steady-state assumption cannot be made, we can derive the reaction rate from (11.5) by integration, resulting in

$$v = V[S]\{1 - \exp[-(k_1[S] + k_2 + k_3)t]\}/(K_m + [S]) \qquad (11.8)$$

Evidently, (11.8) reduces to (11.7) for large values of t.

In the case where $[S]$ does not remain constant, Eqs. (11.7) and (11.8) take some corrections.[14] To estimate the reaction parameters V and K_m, the graphic method has been most commonly employed. Plots of v against $[S]$ that generate a rectangular hyperbola from (11.7) seem quite natural, but in practice are highly unsatisfactory for the above purpose, because it is difficult to estimate two asymptotes. Michaelis and Menten plotted v against $\log[S]$, which gives a symmetrical S-shaped curve with a maximal slope of $0.576V$ at $[S] = K_m$. However, the following three equations from (11.7) have been commonly employed:

$$[S]/v = K_m/V + [S]/V \qquad (11.9)$$

$$v = V - K_m v/[S] \qquad (11.10)$$

$$1/v = 1/V + K_m/(V[S]) \qquad (11.11)$$

All three equations generate linear plots, and V and K_m are respectively the slope and the intercept with the ordinate.

When both substrate and product coexist in a reaction system, the back-reaction in the second step is also possible:

$$E + S \underset{k_2}{\overset{k_1}{\rightleftharpoons}} ES \underset{k_4}{\overset{k_3}{\rightleftharpoons}} E + P \qquad (11.12)$$

The steady-state assumption with respect to ES is now expressed by

$$d[ES]/dt = k_1[E][S] + k_4[E][P] - (k_2 + k_3)[ES] = 0 \qquad (11.13)$$

and the net rate of production of P is then given by

$$v = (k_1 k_3[E_0][S] - k_2 k_4[E_0][P])/(k_2 + k_3 + k_1[S] + k_4[P]) \qquad (11.14)$$

At the initial reaction stage ($[P] = 0$), (11.14) is identical to (11.7), as expected. If we define the parameters V_f, V_r, K_m^s, and K_m^P by

$$V_f = k_3[E_0], \qquad V_r = k_2[E_0]$$

$$(11.15)$$

$$K_m^s = (k_2 + k_3)/k_1, \qquad K_m^P = (k_2 + k_3)/k_4$$

there results

$$v = (V_f[S]/K_m^s - V_r[P]/K_m^P)/(1 + [S]/K_m^s + [P]/K_m^P) \qquad (11.16)$$

This equation is regarded as the general reversible form of the Michaelis–Menten equation. It is quite useful in cases where the relationship between v and $[S]$ is not in line with (11.7).

11.3. Micelle-Catalyzed Reactions

The mechanisms of micelle-catalyzed reactions have been studied not only by analogy with the Michaelis–Menten equation for enzymatic reactions[15] but also from the perspective of volume fractions of the two-part reaction system consisting of the micelles and the intermicellar bulk solutions.[16-19] The kinetics of this reaction has been successfully used only when the micellar concentrations are much higher than the reactant concentrations. However, micellar concentrations near the CMC are often less than the reactant concentrations. In such cases, the distribution of reactants among micelles must be taken into consideration, which is essentially a thermodynamic problem (Chapter 9). *Reactant* is a better technical term than *substrate* for micelle-catalyzed reactions.

The stepwise association of monomeric reactants R with micelles M leads to the distribution of reactants among micelles, and the association reactions can be assumed to be much more rapid than the reaction time (Chapter 4). Hence, the following reaction scheme is considered:

$$M + R \underset{}{\overset{\bar{K}_1}{\rightleftharpoons}} MR_1 \xrightarrow{k_1'} M + P$$

$$k_b \downarrow$$

$$P$$

$$MR_1 + R \underset{}{\overset{\bar{K}_2}{\rightleftharpoons}} MR_2 \xrightarrow{k_2'} MR_1 + P$$

$$\cdots$$

$$MR_{n-1} + R \underset{}{\overset{\bar{K}_n}{\rightleftharpoons}} MR_n \xrightarrow{k_n'} MR_{n-1} + P \qquad (11.17)$$

where k_b is the rate constant of monomer reactant in the bulk phase, and k_i' is that of reactants solubilized in micelles in the form of MR_i. If the reaction of R with X is assumed to be second order, the rate of disappearance of R_t can be rewritten in the form

$$-d[R_t]/dt = k^{app}[R_t][X]$$

$$= (k_b[R] + \sum_{i=1}^{n} k_i'[MR_i])[X] \qquad (11.18)$$

where k^{app} is the apparent rate constant and X may be hydroxide or hydrogen ions, or any reactant other than R. Rearrangement of (11.18) gives the apparent rate constant:

$$k^{app} = (k_b[R] + \sum_{i=1}^{n} k_i'[MR_i])/([R] + \sum_{i=1}^{n} i[MR_i]) \qquad (11.19)$$

For further discussion, we must develop the above equations under reasonable assumptions (Chapter 9). The mass-action law is applicable to the association–dissociation equilibria between micelles and reactants, and the association constant then becomes

$$\bar{K}_i = \vec{k}_{i-1}/\bar{k}_i = [MR_i]/([R][MR_{i-1}]) \qquad (11.20)$$

On the other hand, it is reasonable to assume that k_i' is i times as large as k_1', because, for each reactant, the chance for the reactant R in MR_i to react with X is i times that of MR_1 under the same conditions:

$$k_i' = ik_1' \qquad (11.21)$$

We must next consider the distribution of reactants among micelles. The following are the two most promising distributions.

11.3.1. Poisson Distribution

Let us consider the first equality of (11.20). Assuming that (11.21) is applicable to the dissociation equilibria ($\bar{k}_i = i\bar{k}_1$) and that the rate constant $\vec{k}_i$ of the association of monomer R with MR_i can be made to remain constant regardless of the number of $i(\vec{k}_i = \vec{k}_0)$, we have the following relationship concerning the association constant:

$$\bar{K}_i = \bar{K}_1/i \qquad (11.22)$$

If (11.22) is applied, then $[MR_i]$, $[R_t]$, and $[M_t]$ become the following equations with infinite n:

$$[MR_i] = \bar{K}_1^i[R]^i[M]/i! \tag{11.23}$$

$$[R_t] = [R] + \bar{K}_1[R][M] \exp(\bar{K}_1[R]) \tag{11.24}$$

$$[M_t] = [M] \exp(\bar{K}_1[R]) \tag{11.25}$$

Hence, the average number of reactants per micelle $\bar{R}$ is given by

$$\bar{R} = ([R_t] - [R])/[M_t] = \bar{K}_1[R] \tag{11.26}$$

Thus, the probability that a micelle is associated with i reactants can be written

$$P(i) = \bar{R}^i \exp(-\bar{R})/i! \tag{11.27}$$

This expression is exactly the same as the Poisson distribution. When (11.21) and (11.23) are introduced into (11.19), the expression for the apparent rate constant takes the following form for infinite n:

$$k^{app} = [k_b + k_1'\bar{K}_1[M] \exp(\bar{R})]/[1 + \bar{K}_1[M] \exp(\bar{R})] \tag{11.28}$$

or

$$(k^{app} - k_b)/(k_1' - k^{app}) = \bar{K}_1[M_t] \tag{11.28'}$$

These two equations are the same as those by Menger and Portnoy[20] except for the derivation condition that micellar concentrations are much higher than reactant concentrations.

11.3.2. Gaussian Distribution

The derivation of kinetic equations based on the Gaussian distribution of reactants among micelles is not as definitive as those of the Poisson distribution, because the variable of the Gaussian distribution is continuous, whereas the number of reactants in each micelle is an integer. When $\bar{R}$ is the value at which the distribution becomes maximal (i.e., the mean value of the Gaussian distribution), the normalized Gaussian distribution has the form

$$G(i) = (h/\sqrt{\pi}) \exp[-h^2(i - \bar{R})^2] \tag{11.29}$$

$$\sigma = 1/\sqrt{2}h \tag{11.30}$$

where h is the Gaussian distribution constant relating to the standard deviation σ. Then the concentration of micelles having i reactants $[MR_i]$ can be given by

$$[MR_i] = [M_t]G(i) \tag{11.31}$$

By analogy with (11.19), the apparent rate constant in this case can be written in the form

$$k^{app} = (k_b[R] + [M_t] \int_{-\infty}^{\infty} k'(i)G(i)\,di)/([R] + [M_t] \int_{-\infty}^{\infty} iG(i)\,di) \tag{11.32}$$

In the first place, if k_i' is assumed to be proportional to the number of reactants in a micelle in the same way as in (11.21),

$$k'(i) = (i/\bar{R})k'_{\bar{R}} \tag{11.33}$$

where $k'_{\bar{R}}$ is the rate constant in the micelle associated with $\bar{R}$ reactants. Introducing (11.29) and (11.33) into (11.32) results in

$$k^{app} = \{k_b + k_1'\bar{K}_1\bar{R}[M_t] \exp[h^2(1 - 2\bar{R})]\}/$$
$$\{1 + \bar{K}_1\bar{R}[M_t] \exp[h^2(1 - 2\bar{R})]\} \tag{11.34}$$

or

$$(k^{app} - k_b)/(k_1' - k^{app})$$
$$= \bar{K}_1[M_t]\bar{R} \exp[h^2(1 - 2\bar{R})] \qquad k_1' = k'(1) \tag{11.34'}$$

There remains the question of whether the smooth Gaussian function is applicable to a discrete function. In reality, however, the summation of the Gaussian function in a discrete form can be easily replaced by its integration to within an experimental error.[15]

It is very important to consider which distribution best approximates the real reactant distribution. In order to simulate the distribution, (11.28) and (11.34) are transferred so as to give a plot of $1/(k^{app} - k_b)$ against $1/[M_t]$:
Poisson

$$1/(k^{app} - k_b) = 1/(k_1' - k_b) + 1/[\bar{K}_1(k_1' - k_b)] \times 1/[M_t] \tag{11.35}$$

Gaussian

$$1/(k^{app} - k_b) = 1/(k_1' - k_b)$$
$$+ 1/\{\bar{K}_1(k_1' - k_b)\bar{R}\exp[h^2(1 - 2\bar{R})]\} \times 1/[M_t]   (11.36)$$

Figure 11.2 shows the simulation curves for the two distributions. The plots of $1/(k^{app} - k_b)$ against $1/[M_t]$ give a straight line for the Poisson distribution, wheras those for the Gaussian distribution are concave to the abscissa.

Up to now, the reaction kinetics has been treated assuming monodispersity of the micellar aggregation number. However, polydispersity of the

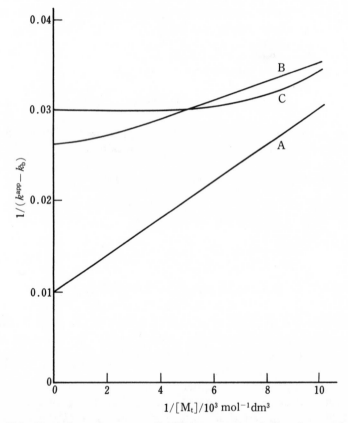

Figure 11.2. Simulation of reactant distribution among micelles from variation of $1/(k^{app} - k_b)$ values with micellar concentration. $\bar{K}_1 = 5 \times 10^3$ mol^{-1} dm^3, $\bar{R} = 5$ at the micellar concentrationof 2×10^{-5} mol dm^{-3}, and $1/(k^{app} - k_b) = 0.01$ at an infinite micellar concentration. A, Poisson distribution; B, Gaussian distribution of $\sigma = 1.5$; C, Gaussian distribution of $\sigma = 0.3\bar{R}$. (Reproduced with permission of the American Chemical Society.)

micelles will affect the reaction kinetics. For example, k_i' variation of the surface area per micelle could result in variation of both the rate constant and the association constant $\bar{K}_i$ of reactant to micelle. However, Eqs. (11.28) and (11.34) are obtained by averaging k_i' and $\bar{K}_i$ over the range of micellar aggregation number.[15]

The reaction of tetranitromethane with hydroxide ion is[21,22]

$$C(NO_2)_4 + OH^- \xrightarrow{k} C(NO_2)_3^- + HNO_3 \qquad (11.37)$$

where k is the second-order rate constant. Figure 11.3 shows the plots obtained using (11.35) and (11.36) for this reaction in a few surfactant systems, and Table 11.1 gives the rate constants obtained from the plots.[23] Because of the excellent linearity of these plots, the distribution of the reactants among micelles can be well approximated by the Poisson distribution.

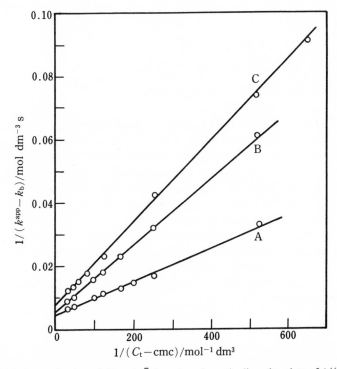

Figure 11.3. Determination of k_1' and $\bar{K}_1/n$ values from the linearity plots of $1/(k^{app} - k_b)$ aginst $1/(C_t - CMC)$. A, hexadecyl trimethylammonium chloride; B, hexadecyl trimethylammonium bromide; C, hexadecyl trimethylammonium nitrate. (Reproduced with permission of the Chemical Society of Japan.)

Table 11.1. Second-Order Rate Constant (k_1') for Nitroform Anion Formation, and the Association Parameter $(\bar{K}_1/n)$ and the Roughly Estimated Association Constant $(\bar{K}_1)$ of Tetranitromethane with Micelles from Micelle Aggregation Number n

Surfactant	$k_1' \times 10^{-2}$ $(dm^3 \cdot mol^{-1} \cdot s^{-1})$	$\bar{K}_1/n \times 10^{-1}$ $(dm^3 \cdot mol^{-1})$	$\bar{K}_1 \times 10^{-2}$ $(dm^3 \cdot mol^{-1})$
Cationic			
$C_{10}H_{21}N^+(CH_3)_3Cl^-$	0.33	1.17	3
$C_{12}H_{25}N^+(CH_3)_3Cl^-$	0.56	3.2	12
$C_{16}H_{33}N^+(CH_3)_3Cl^-$	2.28	8.9	54
$C_{16}H_{33}N^+(CH_3)_3Br^-$	1.76	5.5	44
$C_{16}H_{33}N^+(CH_3)_3NO_3^-$	1.27	6.1	48
$C_{16}H_{33}N^+(CH_3)_3NO_3^-$	1.27	6.1	48
$C_{16}H_{33}N^+C_5H_5Cl^-$	1.52	10.9	71
$C_{16}H_{33}N^+(CH_3)_2CH_2\phi Cl^-$	2.67	18.3	>150
Nonionic			
$C_8H_{17}O(C_2H_4O)_6H$	0.044	2.93	9
$C_{10}H_{21}O(C_2H_4O)_6H$	0.104	1.82	17
$C_{12}H_{25}O(C_2H_4O)_6H$	0.171	1.03	48
$C_{12}H_{25}O(C_2H_4O)_{10}H$	0.67	0.70	12
$C_{12}H_{25}O(C_2H_4O)_{15}H$	0.50	0.83	10
$C_{12}H_{25}O(C_2H_4O)_{20}H$	0.289	0.55	4
$C_{12}H_{25}O(C_2H_4O)_{29}H$	0.142	1.60	9

[a]Reproduced with permission of the Chemical Society of Japan.

The foregoing discussions have regarded micelles as a chemical species, whereas many reports have been based on a pseudophase model and on the partition of reactants between the micellar phase and the intermicellar bulk phase.[17,18,24] Recently, however, the failure of the pseudophase model has been demonstrated even for studies of micellar catalysts.[25,26] Fendler and Fendler[10] summarized data on the reaction rates for a wide variety of micellar systems on the basis of the studies performed up to 1975. During the past 15 years, the focus of interest has shifted from organic reactions to photochemistry in micellar systems (Chapter 12). Nevertheless, the effects of counterions[27,28] and of nonmicellar aggregates of hydrophobic ion[29] on micellar catalysis remain highly instructive as do those on micelle and submicelle formation.

11.4. Inhibition in Micellar Solutions

Compounds that reduce the rate of a reaction are called *inhibitors* (I). Inhibition can be brought about by a wide variety of mechanisms. In enzymatic reactions, inhibitors can be classified as either *irreversible* or *reversible.* The former are catalytic poisons that combine with an enzyme in such a way as to reduce its activity to zero. Many enzymes can be poisoned by trace amounts of heavy metal ions, and for this reason it is common practice to carry out kinetic studies in the presence of complexing agents. Reversible inhibitors—by far the more important class of inhibitor—combine with the enzyme to form a dynamic complex that has lower catalytic activity than the free enzyme. *Competitive inhibitors* compete with the substrate for the substrate binding site on the enzyme, forming an enzyme-inhibitor complex (EI) in place of the enzyme-substrate complex[13]:

$$E + S \underset{}{\overset{K_s}{\rightleftharpoons}} ES \xrightarrow{k_3} E + P$$
$$K_I \Big\Updownarrow {}^{+I}$$
$$EI \tag{11.38}$$

The concentration of this complex is given in a similar manner to (11.2) by the equilibrium constant $K_I = [E][I]/[EI]$, which is called the *inhibition constant.* For sufficiently rapid complex formation, the same procedure that gave (11.4) yields

$$v = V[S]/(K_m^{app} + [S]) \tag{11.39}$$

where

$$K_m^{app} = K_s(1 + [I]/K_I) \tag{11.40}$$

and V and K_s have the same meanings as in Section 11.2. Equation (11.39) is of the same form as the Michaelis-Menten equation.

Inhibition in micelle-catalyzed reactions cannot be handled as easily as inhibition in enzymatic reactions, because reactants and inhibitors do not form complexes of $1:1$ stoichiometry with micelles. As discussed in Chapters 9 and 12, the most promising model for the distribution of solubilizates into micelles is the Poisson distribution, which is employed here. The incorporation of reactants and inhibitors into micelles can be assumed to be independent of each other as long as the amounts incorporated are small and do not change the characteristics of the mother micelle. Then, the denominator of (11.19) remains unchanged. As for the numerator, the MR_i micelle containing no inhibitor experiences the same reaction. If we assume that the reaction is complety stopped for micelles incorporating more than j inhibitor molecules, (11.28) is then transformed to

$$k^{app} = \left[k_b + k_1' \bar{K}_1[M_t]\left(\sum_{j=0}^{j} P(j) \right) \right] \Big/ (1 + \bar{K}_1[M_t]) \tag{11.41}$$

because the Poisson distribution of inhibitors is applicable to each MR_i micelle irrespective of the magnitude of i. When $j = 0$, or when the reaction is totally inhibited in micelles containing any number of inhibitor molecules, (11.41) becomes

$$k^{app} = (k_b + k_1' \bar{K}_1[M_t] - k_1' \bar{K}_1[I])/(1 + \bar{K}_1[M_t]) \qquad (11.42)$$

where the average number of inhibitor molecules per micelle is assumed to be small and the inhibitor molecules are exclusively incorporated into micelles. When the micellar concentration is so great that there are only three kinds of micelle, M, MR_1, and MI_1, we have

$$[M] = ([M_t] - [MR_1])/(1 + K_I[I]) \qquad (11.43)$$

Introducing (11.43) into (11.28) leads to

$$k^{app} = (k_b + k_b K_1[I] + k_1' \bar{K}_1[M_t])/(1 + \bar{K}_1[M_t] + K_1[I]) \qquad (11.44)$$

or

$$(k_1' - k_b)/(k^{app} - k_b) = 1 + 1/(\bar{K}_1[M_t]) + K_1[I]/(\bar{K}_1[M_t]) \qquad (11.44')$$

where $[MI_1]$ is assumed to be negligible compared to $[M]$. Equation (11.44') has been found to be applicable to some inhibited micellar reactions.[30]

Micellar catalysis has no analog for competitive inhibition. Inhibition in micellar systems most often involves an interaction between reactants and micelles. For example, if one ionic reagent is excluded from a counterionic micelle by electrostatic repulsion while the other is concentrated in the micelle, the reaction will be retarded. An increase in the micellar concentration dilutes the reagents in the micelle and thus reduces the apparent reaction rate. An unfavorable location of the reaction site within the micelle can also be a dominant factor resulting in inhibition.[31] As is clear from the above equations and discussions, micelle-catalyzed reactions are essentially different from those catalyzed by enzymes.

References

1. C. A. Bunton, *Prog. Solid State Chem.* 8, 239 (1973).
2. H. Morawetz, *Adv. Catal.* 20, 341 (1969).
3. E.H. Cordes and R. B. Dunlap, *Acc. Chem. Res.* 2, 329 (1969).
4. I. B. Wilson, *Ann. N. Y. Acad. Sci.* 81, 307 (1959).
5. Y. Moroi, *Bull. Chem. Soc. Jpn.* 54, 3265 (1981).

6. T. C. Bruice, in: *The Enzymes* (P. D. Boyer, ed.), 3rd. ed., Vol. 2, p. 217, Academic Press, New York, (1970).
7. E. J. Fendler and J. H. Fendler, *Adv. Phys. Org. Chem. 8*, 271 (1970).
8. I. V. Berezin, K. Martinek, and A. K. Yatsimirski, *Russ. Chem. Rev. 42*, 787 (1973).
9. E. H. Cordes, *Reaction Kinetics in Micelles*, Plenum Press, New York, (1973).
10. J. H. Fendle and E. J. Fendler, *Catalysis in Micellar and Macromolecular Systems*, Academic Press, New York (1975).
11. L. Michaelis and M. L. Menten, *Biochem. Z. 49*, 333 (1913).
12. G. E. Briggs and J. B. S. Haldane, *Biochem. J. 19*, 338 (1925).
13. A. Cornish-Bowden, *Principles of Enzyme Kinetics*, Butterworths, London (1976).
14. K. J. Laidler, *Can. J. Chem. 33*, 1614 (1955).
15. Y. Moroi, *J. Phys. Chem. 84*, 2186 (1980).
16. P. Heitmann, *Eur. J. Biochem. 5*, 305 (1968).
17. A. K. Yatsimirski, K. Martinek, and I. V. Berezin, *Tetrahedron 27*, 2855 (1979); K. Martinek, A. K. Yatsimirski, A. P. Osipov, and I. V. Berezin, *Tetrahedron 29*, 963 (1973).
18. S. J. Dougherty and J. C. Berg, *J. Colloid Interface Sci. 49*, 135 (1974).
19. K. Shiraham, *Bull. Chem. Soc. Jpn. 48*, 2673 (1975).
20. F. M. Menger and C. E. Portnoy, *J. Am. Chem. Soc. 89*, 4698 (1967).
21. D. J. Glover, *J. Phys. Chem. 74*, 21 (1970).
22. J. H. Fendler and R. R. Liechti, *J. Chem. Soc. Perkin Trans. 2, 1972*, 1041.
23. T. Harada, N. Nishikido, Y. Moroi, and R. Matuura, *Bull. Chem. Soc. Jpn. 54*, 2592 (1981).
24. C. A. Bunton, in: *Solution Chemistry of Surfactants* (K. L. Mittal, ed.), Vol. 2, p. 519, Plenum Press, New York (1979).
25. C. A. Bunton, J. Frankson, and L. S. Romsted, *J. Phys. Chem. 84*, 2607 (1980).
26. M. Gonsalves, S. Probst, M. C. Rezende, F. Nome, C. Zucco, and D. Zanette, *J. Phys. Chem. 89*, 1127 (1985).
27. M. Almgren and R. Rydholm, *J. Phys. Chem. 83*, 360 (1979).
28. S. K. Srivastava and S. S. Katiyar, *Int. J. Chem. Kinet. 14*, 1007 (1982).
29. G. Biresaw, C. A. Bunton, C. Quan, and Z.-Y. Yang, *J. Am. Chem. Soc. 106*, 7178 (1984).
30. C. A. Bunton and L. Robinson, *J. Am. Chem. Soc. 90*, 5972 (1968).
31. R. L. Reeves, *J. Am. Chem. Soc. 97*, 6019 (1975).

12

Photochemistry in
Micellar Systems

12.1. Introduction

Photochemistry in micellar systems is a type of micellar catalysis in the sense that the photochemical process takes place in the micellar domain. The outstanding recent progress in micellar photochemistry is well laid out in a number of review articles and papers on photochemical and photophysical processes in micellar assemblies.[1-5] As described in previous chapters, hydrophobic organic solutes solubilize well in the micellar core, whereas the micellar surface controls the concentration of hydrophilic solutes. The electrostatic potential of up to a few hundred millivolts at the surface of ionic micelles is especially effective in attracting or repelling ionic species. Thus, micelles are microscopically heterogeneous and well suited as surfaces for reactions of appropriate reactants.

An important characteristic of reactions in micellar systems is that the micellar concentration can be varied to some extent. In a ususal solvent it is almost impossible to avoid side reactions. In a micellar system, on the other hand, side reactions can be avoided by adjusting the concentrations of reactants and micelles so that most micelles contain just one reactant molecule. The distribution of reactant molecules among micelles thus has a crucial influence on reaction in micellar assemblies, as discussed in Chapter 9.

12.2. Determination of the CMC

Absorption of light by a molecule causes a change in the dipole moment of the molecule. Molecular dipole moments are also influenced by the

medium surrounding the molecule. Thus, molecular absorption spectra
change with the solvent. Because micelles are organized assemblies of
surfactants, properties such as dielectric constant and viscosity vary with
position in the micelle. For example, the dielectric constant is low in the
micellar core and high at the micellar surface. Photochemical processes of
photosensitive probe molecules are very sensitive to the dielectric constant
of the solvent. The precise solubilization site of a probe in a micelle depends
on both the probe and the micelle employed: generally, hydrophobic probes
enter the micellar core and hydrophilic probes localize at the micellar
surface. Thus, micellar structure may be examined by using a probe designed
to localize in a particular micellar zone.

Pyrene monomer fluorescence is a good example of the effects of the
medium on photochemical phenomena. The intensities of the vibronic bands
depend on the solvent, so that the intensity ratio of peak III to I decreases
with increasing dielectric constant of the solvent (Fig. 12.1).[3,6-8] The
maximum wavelength λ_{max} of pyrene carboxyaldehyde increases with an
increase in ε, and, in addition, there is a linear relationship between λ_{max}
and ε over the range $\varepsilon > 10$ (Fig. 12.2).[3] Another example is the dependence
of the absorption maximum of benzophenone[9] on the solvent polarity
parameter (Fig. 12.3).[10-12] Spectral changes of alkylbenzene[13] and ruthenium
bipyridyl ion[14] have also been examined.

It is important to remember that a solubilizate is in association-
dissociation equilibrium with micelles, and that the observed spectrum of
a probe is thus a statistical average over time spent in each location in the
micelle. Determination of the CMC therefore consists of the following two
steps: solubilization of the probe into micelles from an aqueous medium,
and the spectral change due to the different environment. The probe begins
to solubilize at the CMC, when micellization commences, and this event is
marked by an abrupt change in color (Fig. 12.4).[15-19] If the probe concentration is kept constant, the continuing change in color indicates the progress
of solubilization. Most probe molecules are dyes. Their concentration should
be as low as possible so that they do not influence the CMC, as discussed
in Section 4.7. Cationic probes are used for anionic surfactants and anionic
pobes for cationic surfactants. When the presence of the probe may reduce
the CMC, the correct CMC should be determined by extrapolation to a
probe concentration of zero. The dyes used for CMC determination on the
basis of spectral change are pinacyanol chloride,[20-22] rhodamine 6G,[20,22]
erythrosin,[22] sky blue,[20] methyl orange,[23] and TCNQ.[24] Special care must
be taken when pinacyanol chloride is used.[25] In the years since the pyrene
molecule was recognized as a useful fluorescence probe, pyrene and its
derivatives have been employed to determine the CMC of micellar solutions
from the change in the fluorescence spectrum (Fig. 12.5).[26,27]

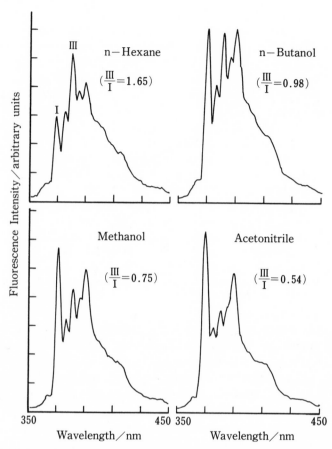

Figure 12.1 Solvent dependence of vibronic band intensities in pyrene monomer fluorescene. [pyrne] = 2 μmol · dm^{-3}; λ_{exit} = 310 nm.[6a] (Reproduced with permission of the American Chemical Society.)

The CMC may also be determined from the rate of energy transfer reactions for which the electrostatic surface potential of an ionic micelle plays an important role in the kinetics.[28]

12.3. Determination of Micellar Aggregation Number

Along with the CMC, the main factor determining the properties of surfactant solution is the micellar aggregation number (the number of surfactant molecules in a micelle). The colligative and light-scattering properties of solutions have been used to determine the micellar aggregation

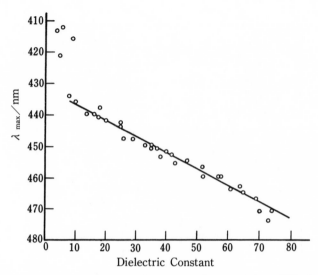

Figure 12.2. Variation of fluorescence maximum λ_{max} of pyrenecarboxyaldehyde with dielectric constant of the medium.[3] (Reproduced with permission of the American Chemical Society.)

number, but these techniques involve the problem of dissociating counterions from the micelle. Photochemical determination completely avoids this problem. Instead, a few reasonable assumptions are employed as to the distribution of probe molecules among micelles and the rate constants of photochemical processes. This method is an outcome of recent progress in photochemistry. Two approaches are used: static[29] and dynamic.[30]

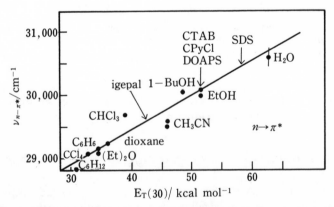

Figure 12.3. Absorption maximum ν (cm^{-1}) of benzophenone plotted against the solvent polarity parameter $E_T(30)$ for $n-\pi^*$ transitions. Surfactant concentration is $0.10 \, mol \cdot dm^{-3}$.[9] (Reproduced with permission of the American Chemical Society.)

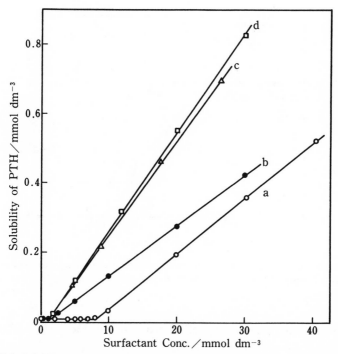

Figure 12.4. Solubility change of phenothiazine (PTH) with surfactant concentrations at 298.15 K. a, sodium dodecyl sulfate (SDS); b, SDS/0.15 mol · dm^{-3} of NaCl; c, manganese (II) dodecyl sulfate; d, zinc(II) dodecyl sulfate.[192] (Reproduced with permission of the American Chemical Society.)

12.3.1. Static Method

The first step in determining the micellar aggregation number by the static method is to select a fluorescence probe P and a quencher Q, both of which are incorporated exclusively into the micellar domain. For anionic surfactant micelles, the typical choices are tris(α, α'-bipyridine)ruthenium ion as P and 9-methylanthracene as Q. In this case, the quenching process takes place only in the micellar region. The concentrations of P and Q and the intensity of the excitation light are all kept small so as not to conflict with the theory given below. In particular, the concentration of P should not exceed one per micelle. The Q concentration is small enough that the Poisson distribution of Q among micelles is applicable: the ratio of micelles [M_i] associated with i molecules of solubilized Q to the total micelles [M_t] is given as

$$[M_i]/[M_t] = \bar{n}_q^i \exp(-\bar{n}_q)/i! \tag{12.1}$$

where $\bar{n}_q$ is an average number of Q molecules per micelle.

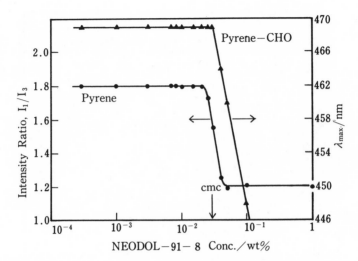

NEODOL−91− 8 Conc./wt%

Figure 12.5. Change in the fluorescence characteristics of pyrene and pyrenecarboxyaldehyde as a function of Neodol 91-8 concentration.[27] (Reproduced with permission of the American Chemical Society.)

There are only three deactivation processes for the excited P*:

$$P^* \xrightarrow{k_f} P + h\nu \tag{12.2}$$

$$P^* \xrightarrow{k_{nr}} P \tag{12.3}$$

$$[P^* \cdots M_i] \xrightarrow{ik_q} [P \cdots M_i]; \qquad i = 1, 2, \cdots \tag{12.4}$$

The first is a radiative decay, the second a nonradiative decay, and the third a quenching decay. According to the form given for (12.4), the kinetics of intramicellar quenching is assumed to be first order and the presence of other quenching molecules in the micelle is assumed not to disturb the process. The constant k_q is the rate constant of the process in the case where one Q and one P coexist in a micelle (Fig. 12.6). The concentration of micelle incorporating one P and iQ is given by the product of $[P_t]$ and (12.1), and the stationary radiation from these micelles becomes

$$d[P^*M_i]/dt = F[P_t] \times [\bar{n}_q^i \exp(-\bar{n}_q)/i!]$$
$$-(k_f + k_{nr} + ik_q)[P^*M_i] = 0 \tag{12.5}$$

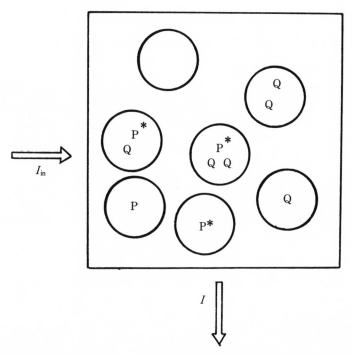

Figure 12.6. Schematic illustration of probe (P) and quenching (Q) molecules solubilized in micelles.

where F is a rate constant of P* formation and [P*] is negligibly small compared with [P_t] owing to weak excitation intensity. The total fluorescence I at a right angle to the incident light by radiative decay then becomes

$$I = f k_f \sum_{i=0}^{\infty} [\text{P*M}_i] \qquad (12.6)$$

where f is an operational constant. Introducing [P*M$_i$] from (12.5) into (12.6) gives

$$I = f F k_f [P_t] \exp(-\bar{n}_q) \times \sum_{i=0}^{\infty} [\bar{n}_q^i / (k_f + k_{nr} + i k_q) i!] \qquad (12.7)$$

In a similar way, the fluorescence intensity I_0 when no Q molecule is present in the system becomes

$$I_0 = f F k_f [P_t] / (k_f + k_{nr}) \qquad (12.8)$$

because from (12.5), $[P^*]$ is

$$[P^*] = F[P_t]/(k_f + k_{nr}) \qquad (12.9)$$

Taking the ratio I/I_0 results in

$$I/I_0 = b \exp(-\bar{n}_q) \times \sum_{i=0}^{\infty} \bar{n}_q^i/[(b + i)i!] \qquad (12.10)$$

where b is given by

$$b = (k_f + k_{nr})/k_q \qquad (12.11)$$

Figure 12.7 shows the relation between I_0/I and $\bar{n}_q$ as a function of parameter b. Particularly when the quenching rate constant is much larger than the rate of deactivation by other mechanisms, (12.10) simplifies to

$$I/I_0 = \exp(-\bar{n}_q) \qquad (12.12)$$

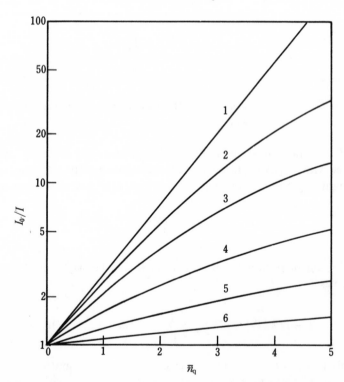

Figure 12.7. Logarithm of the ratio of fluorescence intensities against average number of immobile quenching molecules per micelle. 1, $b = 0$; 2, $b = 0.1$; 3, $b = 0.3$; 4, $b = 1$; 5, $b = 3$; 6, $b = 10$.[29] (Reproduced with permission of Elsevier Science Publishers.)

The average micellar aggregation number $\bar{n}$ is

$$\bar{n} = \sum_n n[\mathrm{M}_n] \Big/ \sum_n [\mathrm{M}_n] = \{C_t - \mathrm{monomer})/[\mathrm{M}_t] \qquad (12.13)$$

and, by combining (12.12) and (12.13), it follows that

$$[\ln(I_0/I)]^{-1} = (1/[\mathrm{Q}]\bar{n}) \times (C_t - \mathrm{monomer}) \qquad (12.14)$$

Plots of the left-hand side of (12.14) against the surfactant concentration at constant Q concentration should be linear, and the slope and the x-intercept are $([\mathrm{Q}]\bar{n})^{-1}$ and CMC, respectively.[31] As is clear from (12.14), it is also possible to obtain the aggregation number at constant surfactant concentration by changing the Q concentration. This method is very valuable for surfactants whose aggregation number changes with total surfactant concentration. In this respect, the present photochemical technique is superior to the conventional light-scattering technique, which requires measurements over a wide range of concentration.

The method has been criticized on the grounds that it does not give an inherent aggregation number because of the coexistence of probe molecules in micelles. However, the aggregation numbers determined by the luminescence quenching method are almost identical to those determined by the light-scattering method (60 ± 2 for SDS, for example).[31] Other reports concern the effects of salt[32] and polar organic additives[33] on SDS or cetyltrimethylammonium chloride micelles,[34] aerosol OT microemulsions,[35] SDS/zinc dodecyl sulfate mixed micelles,[36] and nonionic (Triton) micelles with varying ethylene oxide units.[37] For SDS, an extensive investigation has been reported on the interaction between *tris*(α, α'-bipyridine)-ruthenium ion and dodecyl sulfate ion.[38]

12.3.2. Dynamic Method.

If the lifetime of the fluorescent probe P* is long enough to allow time-resolved fluorescence measurements, the micellar aggregation number can be determined from the decay curve. By using a cutoff of the incident light ($F = 0$), (12.5) can thus be written as

$$d[\mathrm{P}^*\mathrm{M}_i]/dt = -(k_f + k_{nr} + ik_q)[\mathrm{P}^*\mathrm{M}_i] \qquad (12.15)$$

Integration of (12.15) with respect to time and a subsequent summation of i leads to

$$[\mathrm{P}^*(t)] = [\mathrm{P}^*(0)] \times \exp\{-(k_f + k_{nr})t + \bar{n}_q[\exp(-k_q t) - 1]\} \qquad (12.16)$$

where the Poisson distribution of Q and the following relation are employed:

$$\sum_{i=0}^{\infty} [\bar{n}_q^i \exp(-ik_q t)/i!] = \exp[\bar{n}_q \exp(-k_q t)] \tag{12.17}$$

If (12.16) is introduced into (12.6) and $\log[I/I_0]$ is graphed against time, a linear plot is obtained for which k_q is the initial decay, $k_f + k_{nr}$ is the slope for a longer time scale, and $\bar{n}_q$ is the x-intercept.

Another dynamic method is to use a time-resolved fluorescence measurement[30] in which the kinetics of intramicellar excimer formation and dissociation are taken into consideration. The excimer is a dimer made of activated and nonactivated molecules. The excimer of pyrene in micelles has been investigated in detail,[8,39,40] and pyrene has been employed in most cases.

Micelles incorporating n probe molecules (MP_n) obey the Poisson distribution with respect to n, and the following are the four fundamental intramicellar photochemical processes:

$$MP_n \xrightarrow{P(n)I_a} MP^*P_{n-1} \tag{12.18}$$

$$MP^*P_{n-1} \xrightarrow{k_1} MP_n \tag{12.19}$$

$$MP^*P_{n-1} \underset{k_{-E}}{\overset{(n-1)k_E}{\rightleftharpoons}} MP_2^*P_{n-2} \tag{12.20}$$

$$MP_2^*P_{n-2} \xrightarrow{k_1'} MP_n \tag{12.21}$$

The first is an excitation of P, the second a radiative transition, the third excimer formation and dissociation, and the fourth a transition decay of the excimer. In every case, the kinetics is assumed to be first order, $P(n)$ is a distribution probability from (12.1), I_a is the intensity of incident light, and k_1, k_E, k_{-E}, and k_1' are the rate constants of corresponding reactions. From the above processes, the following two rate equations are derived concerning MP^*P_{n-1} and MP^*P_{n-2}:

$$d[MP^*P_{n-1}]/dt = P(n)I_a - [k_1 + (n-1)k_E][MP^*P_{n-1}]$$
$$+ k_{-E}[MP_2^*P_{n-2}] \tag{12.22}$$

$$d[MP_2^*P_{n-2}]/dt = (n-1)k_E[MP^*P_{n-1}]$$
$$- (k_1' + k_{-E})[MP_2^*P_{n-2}] \tag{12.23}$$

The time dependence of the fluorescence intensity from the decay of excited monomer after pulse excitation may be simplified from (12.22) by assuming that $I_a = 0$ and $k_1' \gg k_{-E}$:

$$d[MP^*P_{n-1}]/dt = -[k_1 + (n-1)k_E] \times [MP^*P_{n-1}] \qquad (12.24)$$

The time dependence of fluorescence intensity obtained by summarizing the above solution with respect to n is then given by

$$\ln[I(t)/I(0)] = -k_1 t + \bar{n}[\exp(-k_E t) - 1] \qquad (12.25)$$

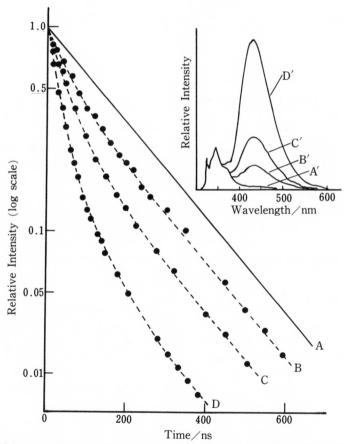

Figure 12.8. Pyrene in a cetyltrimethylammonium chloride (CTAC) micelle at [CTAC] = $0.010\ \text{mol} \cdot \text{dm}^{-3}$. A, [pyrene] = $7.5 \times 10^{-6}\ \text{mol} \cdot \text{dm}^{-3}$; B, $5.2 \times 10^{-5}\ \text{mol} \cdot \text{dm}^{-3}$; C, $1.0 \times 10^{-4}\ \text{mol} \cdot \text{dm}^{-3}$; D, $2.08 \times 10^{-4}\ \text{mol} \cdot \text{dm}^{-3}$; ●, experimental points from transition experiment for monomer decay. Inset: longer wavelength emission from the pyreme excimer obtained by steady-state experiment normalized to emission for A.[4]

Table 12.1. CMC and Micelle Aggregation Number ($\bar{n}$) of SDS and n-Alkylsulfonic Acids[a]

Surfactant	CMC[b] (mmol $\cdot$ dm^{-3})	CMC[c] (mmol $\cdot$ dm^{-3})	$\bar{n}^c$	$\bar{n}^d$
SDS	8.2	5.7	59	64
$C_{12}H_{25}SO_3H$	7.8	7.5	58	63
$C_{14}H_{29}SO_3H$	2.2	0.32	85	82
$C_{16}H_{33}SO_3H$	0.59	-0.84	101	107

[a] Reproduced with permission of Academic Press.
[b] CMC determined by electrical conductivity.
[c] Values determined from steady-state fluorescence data.
[d] Values determined from fluorescence decay measurements.

where the Taylor series of (12.17) is also employed. Thus, the time-resolved fluorescence intensity can be used to calculate the rate constants k_1 and k_E, and the average of probe molecules per micelle $\bar{n}$. Equations (12.16) and (12.25) are fundamentally the same. To perform this determination, a wavelength due to monomer decay must be selected, because the radiative transitions from monomer and excimer take place at the same time.

Figure 12.8 shows examples for micelles of cetyltrimethylammonium chloride,[30] and Table 12.1 gives micellar aggregation numbers determined by both static and dynamic methods for SDS and alkylsulfonic acids.[41] The results are in good agreement within the range of experimental error.

12.4. Kinetics of Redox Reactions

One of the most important characteristics of ionic micelles is their electrostatic potential (up to hundreds of millivolts at the micellar surface) and their resulting ability to select specific counterion species. The potential also depends on the counterion, so the above two quantities are not necessarily independent. They have a crucial effect on electron transfer reactions in micellar systems.

When the following photo-redox reaction takes place between an electron acceptor A and an electron donor D in a micellar system:

$$A + D \underset{\Delta}{\overset{h\nu}{\rightleftharpoons}} A^- + D^+ \tag{12.26}$$

the effective conversion of light energy to chemical energy depends absolutely on whether the products A^- and D^+ formed by the light energy can be made to undertake another chemical reaction before the rapid back reaction dissipates the energy as heat (Δ). A promising approach is to employ charged surfactant aggregates, such as micelles or vesicles, as a reaction medium. The role of these assemblies is to (1) solubilize the photoactive species in the micellar domain and (2) provide an ultrathin electrostatic barrier at the micellar surface,[42] by means of which kinetic control of the forward and back reactions of (12.26) becomes feasible.[43]

Phenothiazine and its derivatives are photochemically interesting, and their triplet state and cation radicals have been extensively investigated.[44-46] Electron transfer reactions using these agents as electron donors have been studied in both homogeneous[47,48] and micellar solutions.[29,49-52] Figure 12.9 shows the spectra of 10-methylphenothiazine (MPTH) solubilized in anionic micelles with cationic metallic counterions (MPTH^T for Na$^+$ and MPTH$^+$ for Cu^{2+}) immediately after a laser pulse.[49] From time-resolved analysis over a brief interval, the electron transfer takes place via MPTH^T. However, over a long time scale, the reaction can be expressed as

$$MPTH + Cu^{2+} \xrightarrow{h\nu} MPTH^+ + Cu^+ \qquad (12.27)$$

The MPTH$^+$ thus produced cannot escape from the mother micelle because of its hydrophobic interaction with the micelle and the micelle's negative

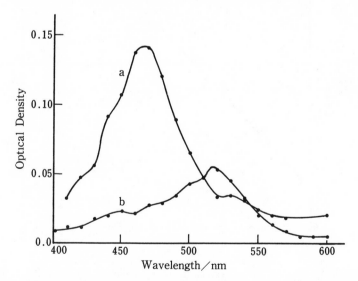

Figure 12.9. Absorption spectra immediately after a laser pulse. a, 1×10^{-4} mol · dm^{-3} MPTH in 5×10^{-2} mol · dm^{-3} SDS solution; b, 1×10^{-4} mol · dm^{-3} MPTH in 2×10^{-2} mol · dm^{-3} copper(II) dodecyl sulfate.[49] (Reproduced with permission of the American Chemical Society.)

electrostatic potential. On the other hand, reduced Cu^+ is forced to move far from the mother micelle, and the electron transfer back reaction is thus second order (Fig. 12.10). If Cu^+ remained in the micelle, the back reaction would be first order because of the decay of D^+-A^- pairs in the micelle.

The following example is a versatile electrokinetic mechanism that can be applied to a variety of electron transfer reactions.[50] The micellar systems is composed of anionic micelles incorporating MPTH, in which the counterion Cu^{2+} is replaced by Eu^{3+}. If Eu^{3+} is the only counterion, the Krafft temperature range (or MTR) of the surfactant is too high to use, so micelles with mixed Eu^{3+} and Na^+ counterions are used instead. The electron transfer

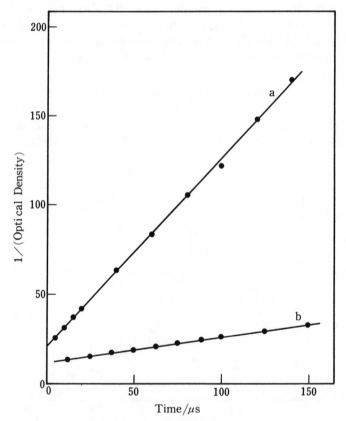

Figure 12.10. Kinetics of MPTH cation decay. a, 1×10^{-4} mol $\cdot$ dm^{-3} MPTH in 2×10^{-2} mol $\cdot$ dm^{-3} copper(II) dodecyl sulfate; b, 3×10^{-4} mol $\cdot$ dm^{-3} MPTH in 5×10^{-3} mol $\cdot$ dm^{-3} CuSO$_4$ in ethanol/water (1/2 v/v) mixed solvent solution.[49] (Reproduced with permission of the American Chemical Society.)

reaction is similar to that of Cu^{2+}:

$$MPTH + Eu^{3+} \xrightarrow{h\nu} MPTH^+ + Eu^{2+} \qquad (12.28)$$

The succeeding electron transfer back reaction between $MPTH^+$ and Eu^{2+} consists of three elementary processes (Fig. 12.11): (1) first-order back reaction of D^+-A^- ion pairs in a micelle (k_b), (2) dissociation (k_d) and association (k_a) of A^- with micelles, and (3) hopping transfer of A^- among micelles through the overlapping ionic atmosphere formed by close approach of micelles (k_h). The time dependence of the respective $[D^+A^-]$, $[D^+]$, $[MA^-]$, and $[A^-]$ are

$$d[D^+A^-]/dt = -(k_b + k_d + k_h[M])[D^+A^-] + k_h[D^+][MA^-]$$
$$+k_a[D^+][A^-] \qquad (12.29)$$

$$d[D^+]/dt = (k_d + k_h[M])[D^+A^-] - k_a[D^+][A^-]$$
$$-k_h[D^+][MA^-] \qquad (12.30)$$

$$d[MA^-]/dt = k_h[M][D^+A^-] - k_d[MA^-] + k_a[M][A^-]$$
$$-k_h[D^+][MA^-] \qquad (12.31)$$

$$d[A^-]/dt = k_d([D^+A^-] + [MA^-]) - k_a([M][A^-]$$
$$+[D^+][A^-]) \qquad (12.32)$$

where $[M]$ refers to empty and D-containing micelles, $[D^+]$ to micelles with associated D^+, $[MA^-]$ to micelles with associated A^-, $[D^+A^-]$ to micelles with associated D^+ and A^-, and $[A^-]$ to free A^- in the intermicellar bulk phase. The observable parameter is the transient optical density of the sum of $[D^+]$ an $[D^+A^-]$:

$$[D_t^+] = [D^+] + [D^+A^-] \qquad (12.33)$$

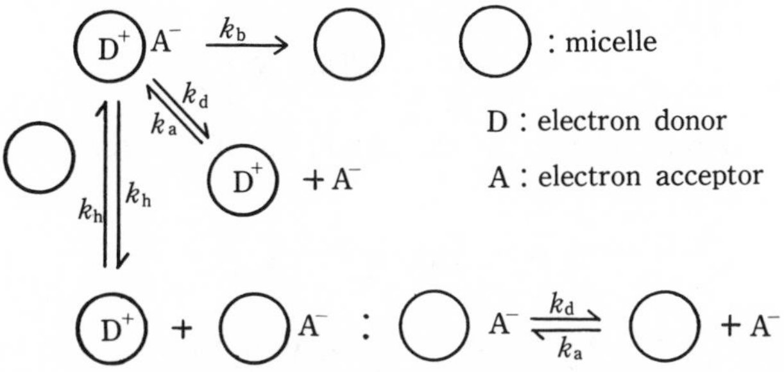

Figure 12.11. Schematic illustration of the elementary processes contributing to the back-transfer of an electron from a reduced acceptor to an oxidized donor in a micellar solution.[50] (Reproduced with permission of the American Chemical Society.)

When the decay of $[D_t^+]$ over a brief interval is the point of discussion, the second-order terms of (12.29) and (12.30) are negligible, and there results

$$d[D^+A^-]/dt = -(k_b + k_d + k_h[M])[D^+A^-] \tag{12.34}$$

$$d[D^+]/dt = (k_d + k_h[M])[D^+A^-] \tag{12.35}$$

Integration of (12.34) with respect to time gives

$$[D^+A^-] = C^0 \exp[-(k_b + k_d + k_h[M])t] \tag{12.36}$$

and substitution of the result of $[D^+A^-]$ into (12.35) leads to

$$[D^+] = [C^0(k_d + k_h[M])/(k_b + k_d + k_h[M])]$$
$$\times\{1 - \exp[-(k_b + k_d + k_h[M])t]\} \tag{12.37}$$

where C^0 is the concentration of D^+A^- immediately after excitation. The sum of (12.36) and (12.37) results in

$$[D_t^+] = [C^0/(k_b + k_d + k_h[M])]$$
$$\times\{k_d + k_h[M] + k_b \exp[-(k_b + k_d + k_h[M])t]\} \tag{12.38}$$

The experimental parameter $[D_t^+]$ approaches a constant value after initial first-order decay. Over a longer interval, on the other hand, the system reaches a steady state for the concentration of D^+A^-:

$$d[D^+A^-]/dt \simeq 0 \tag{12.39}$$

The partitioning of reduced acceptors between the aqueous bulk phase and the micellar domain is also stationary:

$$k_d[MA^-] \simeq k_a[M][A^-] \tag{12.40}$$

An additional relationship for the condition of electroneutrality of the solution is

$$[D^+] = [MA^-] + [A^-] \tag{12.41}$$

From (12.29), (1.39), (12.40), and (12.41), all transient concentrations can be expressed in terms of $[D^+]$:

$$[A^-] = [D^+]/(1 + k_a[M]/k_d) \tag{12.42}$$

$$[MA^-] = (k_a[M]/k_d) \times [[D^+]/(1 + k_a[M]/k_d)] \qquad (12.43)$$

$$[D^+A^-] = [(1 + k_h[M]/k_d)/(1 + k_a[M]/k_d)]$$
$$\times[k_a[D^+]^2/(k_b + k_d + k_n[M])] \qquad (12.44)$$

By inserting the above three concentrations into (12.29) and (12.30), there results

$$d[D^+]/dt = -k_{obs}[D^+]^2 \qquad (12.45)$$

where k_{obs} is the second-order rate constant expressed as

$$k_{obs} = [k_a k_b/(k_b + k_d + k_h[M])]$$
$$\times[(1 + k_h[M]/k_d)/(1 + k_a[M]/k_d)] \qquad (12.46)$$

Over the longer interval, $[D^+A^-]$ is negligible compared to $[D^+]$, and (12.45) becomes:

$$d[D_t^+]/dt = -k_{obs}[D_t^+]^2 \qquad (12.47)$$

The next problem is to determine the four rate constants from the two parameters obtainable by experiment. One is the rate constant k of the first-order decay in the short time domain

$$k = k_b + k_d + k_h[M] \qquad (12.48)$$

and the other is the rate constant of the second-order decay in the longer time domain, k_{obs}. Plotting the rate constant k obtained by applying (12.38) to an initial rapid decay against micellar concentration, one obtains k_h. The k value increases with increaing micellar concentration, which is the reasoning behind the hopping reaction of A^-. From differentiation of (12.38) with respect to time, the slope at time zero becomes

$$-d[D_t^+]/dt|_{t=0} = C^0 k_b \qquad (12.49)$$

Because C^0 is obtainable from the optical density at time zero (A_0) and the molar extinction coefficient of D^+, k_b can be determined by the initial slope, after which k_d can be automatically evaluated.

Another approach to determining the rate constants is to use the following equation resulting from (12.38):

$$1/(A_0/A_p - 1) = k_d/k_b + (k_h/k_b)[M] \qquad (12.50)$$

where A_p is the optical density of the plateau domain after the initial decay. Thus, from the linear plots of the left-hand side against the micellar concentration, the values of k_d/k_b and k_h/k_b follow from the intercept and the

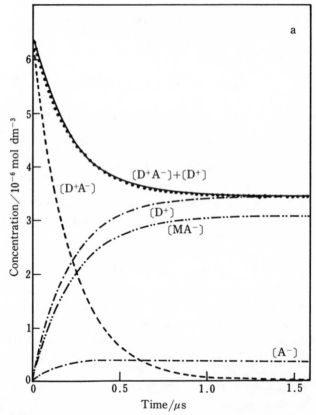

Figure 12.12. Time dependence of the transient concentrations for chemical species produced by a light-initiated redox reaction in an SDS/europium(III) decyl sulfate system, $k_b = 2.1 \times 10^6 \text{ s}^{-1}$, $k_d = 0.45 \times 10^6 \text{ s}^{-1}$, $k_h = 2.6 \times 10^9 \text{ mol}^{-1} \cdot \text{dm}^3 \cdot \text{s}^{-1}$, $k_a = 5.2 \times 10^9 \text{ mol}^{-1} \cdot \text{dm}^3 \cdot \text{s}^{-1}$, $[M] = 7.6 \times 10^{-4} \text{ mol} \cdot \text{dm}^{-3}$, $[D^+]_{t=0} = 6.42 \times 10^{-6} \text{ mol} \cdot \text{dm}^{-3}$. a, short-time decay; b, long-time decay; ——, experimental curve.[50] (a) Time/μs; (b) time/100 μs. (Reproduced with permission of the American Chemical Society.)

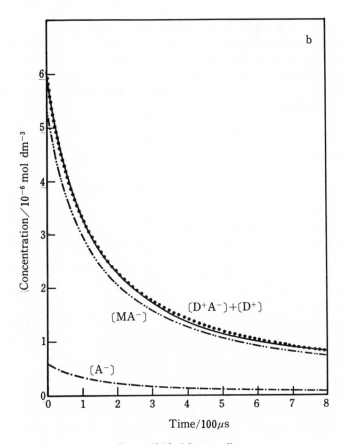

Figure 12.12. (Continued)

slope of the line, respectively. The decay of $[D_t^+]$ in the longer time domain is second order, and the decay constant k_{obs} decreases with increasing micellar concentration, as expected from (12.46). The k_a value can then be finally evaluated from the values of k_{obs}, k_b, k_d, and k_h. Figure 12.12 shows the concentration change with time for each chemical species.

Other examples of electron transfer reactions in surfactant assemblies are those between pyrene and dimethylaniline in micelles,[53] between viologen derivative and zinc porphyrin as an electron relay,[54] and between chlorophyll *a* and methylviologen in microemulsions[55]; the photoinduced reduction of duroquinone by zinc porphyrin in micellar solution[56]; the photoinduced redox reaction of proflavine in aqueous and micellar solution[57]; retardation of back reactions in micellar systems[58]; light-driven electron transfer from tetrathiafulvalene to porphyrin and *tris*(α, α'-bipyridine) rhuthenium ion in micellar and microemulsion solutions[59]; the photoinduced electron transfer from pyrene to cupric ions in anionic micelle[60]; and the effect of ethylenediaminetetraacetate on electron transfer from aqueous colloidal cadmium sulfate to methylviologen.[61] All of these examples are characterized by reactions and reaction rates that can be observed only in surfactant assemblies.

References

1. M. Grätzel and J. K. Thomas, *Modern Fluorescence Spectroscopy*, Vol. 2, Chapter 4, Plenum Press, New York (1976).
2. N. J. Turro, M. Grätzel, and A. M. Braun, *Angew. Chem. Int. Ed. Engl.* 19, 675 (1980).
3. J. K. Thomas, *Chem. Rev. 80*, 283 (1980).
4. L. A. Singer, in: *Solution Behavior of Surfactants* (K. A. Mittal and E. J. Fendler, eds.), Vol. 1, p. 73, Plenum Press, New York (1982).
5. (a) N. J. Turro, G. S. Cox, and M. A. Paczkowski, *Topics in Current Chemistry*, Vol. 129, p. 57, Springer-Verlag, Berlin (1985). (b) K. Kano and T. Matsuo, *Bull. Chem. Soc. Jpn.* 47, 2836 (1974). (c) L. J. Heidt and A. F. McMillan, *J. Am. Chem. Soc. 76*, 2135 (1954). (d) M. Yasuda, T. Yamashita, and K. Shima, *J. Org. Chem. 52*, 753 (1987).
6. (a) K. Kalyanasundram and J. K. Thomas, *J. Am. Chem. Soc. 99*, 2039 (1977). (b) M. Almgren, F. Grieser, and J. K. Thomas, *J. Am. Chem. Soc. 102*, 3188 (1980). (c) S. J. Gregoritch and J. K. Thomas, *J. Phys. Chem. 84*, 1491 (1980).
7. A. Nakajima, *Bull. Chem. Soc. Jpn. 44*, 3272 (1971).
8. R. C. Dorrance and T. F. Hunter, *J. Chem. Soc. Faraday Trans. 1, 73*, 1891 (1977).
9. J. H. Fendler, E. J. Fendler, G. A. Infante, P.-S. Shih, and L. K. Patterson, *J. Am. Chem. Soc. 97*, 89 (1975).
10. K. Dimroth, C. Reichardt, T. Siepmann, and F. Bohlman, *Justus Liebigs Ann. Chem. 661*, 1 (1963).
11. K. Dimroth, C. Reichardt, and T. Siepmann, *Justus Liebigs Ann. Chem. 727*, 93 (1969).
12. E. M. Kosower, *An Introduction to Physical Organic Chemistry*, Wiley, New York (1968).
13. P. Mukerjee, in: *Solution Chemistry of Surfactants* (K. L. Mittal, ed.), Vol. 1, p. 153, Plenum Press, New York (1979).
14. D. Meisel, M. S. Matheson, and J. Rabani, *J. Am. Chem. Soc. 100*, 117 (1978).
15. I. M. Kolthoff and W. Stricks, *J. Phys. Colloid Chem. 52*, 915 (1948).
16. I. M. Kolthoff and W. Stricks, *J. Phys. Colloid Chem. 53*, 424 (1949).
17. R. J. Williams, J. N. Phillips, and K. J. Mysels, *Trans. Faraday Soc. 51*, 728 (1955).
18. T. Nakagawa, K. Kuriyama, S. Inaba, and K. Tohori, *Nippon Kagaku Kaishi 77*, 1563 (1956).
19. (a) Y. Moroi, K. Sato, and R. Matuura, *J. Phys. Chem. 86*, 2463 (1982). (b) Y. Moroi, H. Norma, and R. Matuura, *J. Phys. Chem. 87*, 872 (1983).
20. M. L. Corrin and W. D. Harkins, *J. Am. Chem. Soc. 69*, 679 (1947).
21. S. H. Herzfeld, *J. Phys. Chem. 56*, 953 (1952).
22. T. Nakagawa, K. Tohori, and K. Kuriyama, *Nippon Kagaku Kaishi 77*, 1684 (1956).
23. C. F. Hiskey and T. A. Downer, *J. Phys. Chem. 58*, 835 (1954).
24. (a) S. Muto, K. Deguchi, K. Kobayashi, E. Kaneko, and K. Jeguro, *J. Colloid Interface Sci. 33*, 475 (1970). (b) K. Deguchi and K. Megro, *J. Colloid Interface Sci. 38*, 596 (1972).
25. P. Mukerjee and K. J. Mysels, *J. Am. Chem. Soc. 77*, 2937 (1955).

26. A. Nakajima, *Bull. Chem. Soc. Jpn. 50,* 2473 (1977).
27. K. P. Ananthapadmanabhan, E. D. Goddard, N. J. Turro, and P. L. Kuo, *Langmuir 1,* 352 (1985).
28. M. Maestri, P. P. Infelta, and M. Grätzel, *J. Chem. Phys. 69,* 1522 (1978).
29. P. P. Infelta, *Chem. Phys. Lett. 61,* 88 (1979).
30. S. S. Atik, M. Nam, and L. A. Singer, *Chem. Phys. Lett. 67,* 75 (1979).
31. N. J. Turo and A. Yekta, *J. Am. Chem. Soc. 100,* 5951 (1978).
32. (a) P. Lianos and R. Zana, *J. Phys. Chem. 84,* 3339 (1980). (b) M. Almgren and J.-E. Lofroth, *J. Colloid Interface Sci. 81,* 486 (1981).
33. M. Almgren and S. Swarup, *J. Colloid Interface Sci. 91,* 256 (1983).
34. E. Roelants, E. Gelade, M. Van Der Auweraer, Y. Croonen, and F. C. De Schryver, *J. Colloid Interface Sci. 96,* 288 (1983).
35. A. M. Ganz and B. E. Boeger, *J. Colloid Interface Sci. 109,* 504 (1986).
36. G. G. Warr, F. Grieser, and D. F. Evans, *J. Chem. Soc. Faraday Trans. 1, 82,* 1829 (1986).
37. N. J. Turro and P.-L. Kuo, *Langmuir 1,* 170 (1985).
38. J. H. Baxendale and M. A. J. Rodgers, *J. Phys. Chem. 86,* 4906 (1982).
39. R. C. Dorrance and T. F. Hunter, *J. Chem. Soc. Faraday Trans. 1 68,* 1312 (1972).
40. R. C. Dorrance and T. F. Hunter, *J. Chem. Soc. Faraday Trans. 1 70,* 1572 (1974).
41. Y. Moroi, R. Humphry-Baker, and M. Grätzel, *J. Colloid Interface Sci. 119,* 588 (1987).
42. (a) S. C. Wallace, M. Grätzel, and J. K. Thomas, *Chem. Phys. Lett. 23,* 359 (1973). (b) M. Grätzel and J. K. Thomas, *J. Phys. Chem. 78,* 2248 (1974). (c) S. A. Alkaitis and M. Grätzel, *J. Phys. Chem. 98,* 3549 (1976). (d) C. Wolff and M. Grätzel, *Chem. Phys. Lett. 52,* 542 (1977).
43. (a) M. Calvin, *Photochem. Photobiol. 23,* 425 (1976). (b) G. Porter and M. D. Archer, *Interdiscip. Sci. Rev. 1,* 119 (1976). (c) M. Grätzel, in: *Micellization, Solubilization, and Microemulsions* (K. L. Mittal, ed.), Vol. 2, p. 531, Plenum Press, New York (1977).
44. D. C. Borg and G. C. Citzias, *Proc. Natl. Acad. Sci. U.S.A. 48,* 617 (1962).
45. N. H. Litt and J. Rzdovic, *J. Phys. Chem. 78,* 1750 (1974).
46. S. A. Alkaitis, M. Grätzel, and A. Henglein, *Ber. Bunsenges. Phys. Chem. 79,* 541 (1975).
47. S. A. Alkaitis, G. Beck, and M. Grätzel, *J. Am. Chem. Soc. 97,* 5723 (1975).
48. Y. Kawanishi, N. Kitamura, and S. Tazuke, *J. Phys. Chem. 90,* 6034 (1986).
49. Y. Moroi, A. Braun, and M. Grätzel, *J. Am. Chem. Soc. 101,* 567 (1979).
50. Y. Moroi, P. P. Infelta, and M. Grätzel, *J. Am. Chem. Soc. 101,* 573 (1979).
51. C. Minero, E. Pramauro, E. Pelizzetti, and D. Meisel, *J. Phys. Chem. 87,* 399 (1983).
52. Y. Moroi, *Bull. Chem. Soc. Jpn. 54,* 3265 (1981).
53. J. K. Thomas and M. Almgren, in: *Solution Chemistry of Surfactats* (K. L. Mittel, ed.), Vol. 2, p. 559, Plenum Press, New York (1979).
54. M. Grätzel, *Faraday Discuss. Chem. Soc. 70,* 359 (1980).
55. M.-P. Pileni and M. Grätzel, *J. Phys. Chem. 84,* 2402 (1980).
56. J. Kiwi and M. Grätzel, *J. Phys. Chem. 84,* 1503 (1980).
57. M.-P. Pileni and M. Grätzel, *J. Phys. Chem. 84,* 1822 (1980).
58. P.-A. Brugger, M. Grätzel, T. Guarr, and G. McLendon, *J. Phys. Chem. 86,* 944 (1982).
59. C. K. Grätzel and M. Grätzel, *J. Phys. Chem. 86,* 2710 (1982).
60. T. Nakamura, A. Kira, and M. Imamura, *J. Phys. Chem. 88,* 3435 (1984).
61. J. Kuczynski and J. K. Thomas, *Langmuir 1,* 158 (1985).

13

Interactions between
Amphiphiles and Polymers

13.1. Introduction

Starting with the description by Bull and Neurath of an interaction between SDS and egg albumin,[1] the interaction between amphiphiles and polymers has been widely investigated. Both naturally occurring proteins and starches and synthetic polymers, including polyelectrolytes such as polyacrylic acid and nonionic polymers such as polyethylene oxide, have been investigated. Amphiphiles, on the other hand, may be cationic, anionic, or nonionic. The interaction thus depends on the combinations of amphiphile and polymer.

Study of interactions between synthetic polymers and surfactants began with the observation that the water-insoluble polymer polyvinylacetic acid becomes water-soluble in the presence of excess anionic surfactant.[2] Similar findings followed.[3-5] Moreover, the aqueous solubility of anionic surfactants at temperatures below the micelle temperature range (MTR) increases greatly in the presence of a water-soluble polymer.[6] These two phenomena result from the interaction between polymer and surfactant. The surfactants bind to the polymer, forming a type of complex with a polyvalent charge that behaves in aqueous solution very much like a polyelectrolyte. This similarity can be verified by such measurements as viscosity,[7,8] agglomeration,[9] electrophoresis,[10,11] electric potential difference,[12] dialysis,[13] and counterion dissociation of surfactant from the complex.[14] In general, the strong interactions have been observed between anionic surfactants and nonionic polymers.

Many reports and review articles[15-20] have appeared on the interaction between proteins and surfactants, ever since surfactants were found to be strong denaturants of water-soluble proteins.[21] Much information on proteins is still derived from studies of their interaction with amphiphiles.

Recent interest focuses on the determination of protein molecular weight by electrophoresis with SDS,[22,23] on the separation and purification of proteins from living tissue by solubilization with surfactants,[24,25] and on solution behavior of proteins.[26-30] For convenience, most of these studies use anionic surfactants, for which abundant reference data are available, and serum albumins, which bind readily with a variety of species. The serum albumins function as carrier proteins for nutrients, metabolites, and drugs in the bloodstream. Only a few investigations have employed cationic surfactants.[31-35]

13.2. Analytic Models for Binding

The interaction between polymers (P) and amphiphiles (A) can be concretely represented by a binding isotherm, and analysis of the isotherm yields an estimate of the change of the thermodynamic variables upon binding. Many mathematical descriptions of binding have appeared but only a few of the more common models are introduced in this section.

The amount of bound amphiphile (C_b) is obtained by subtracting the final equilibrium concentration (C_f) from the total concentration (C_t). Thus, the number of bound molecules per molecule of protein (r) becomes

$$r = C_b/C_p = (C_t - C_f)/C_p \tag{13.1}$$

where C_p is the protein concentration. The variation of r with C_f is very important for understanding the mechanism by which amphiphiles bind to proteins.

Scatchard derived a binding equation[36] for the albumin molecule which is very similar to the Langmuir equation.[37] The following assumptions were used: (1) Binding sites are localized and independent (binding at one site does not affect binding at another site). (2) The binding sites can be divided into several kinds, each with homogeneous binding site properties. When the number of binding sites of one kind is n, the binding isotherm is identical with the Langmuir equation and is expressed by

$$r = nKC_f/(1 + KC_f) \tag{13.2}$$

where K is the equilibrium binding constant. The K and n values can be determined by plotting r/C_f against r, as is clear from the rearrangement of (13.2):

$$r/C_f = nK - Kr \tag{13.3}$$

For m kinds of binding sites, (13.2) becomes

$$r = \sum_{i=1}^{m} n_i K_i C_f / (1 + K_i C_f) \tag{13.4}$$

In reality binding sites are not always independent. For example, binding may be independent but cooperative, or amphiphile binding may cause a conformational change that induces new binding sites, or multiple amino acid residues may contribute to the binding. Thus their binding in general should be treated by the following stepwise binding equilibria:

$$P + A \overset{K_1}{\rightleftharpoons} PA_1$$

$$PA_1 + A \overset{K_2}{\rightleftharpoons} PA_2 \tag{13.5}$$

$$\cdots$$

$$PA_{n-1} + \overset{K_n}{\rightleftharpoons} PA_n$$

where K_i is the stepwise binding constant expressed by

$$K_i = [PA_i] / ([PA_{i-1}]C_f) \qquad C_f = [A] \tag{13.6}$$

and n is the maximum number of binding sites. The r value is then expressed by

$$r = \sum_{i=1}^{n} i[PA_i] \Big/ \sum_{i=0}^{n} [PA_i] = \sum_{i=1}^{n} i \left(\prod_{i=1}^{i} K_j \right) C_f^i \Big/ \left[1 + \sum_{i=1}^{n} \left(\prod_{j=1}^{i} K_j \right) C_f^i \right] \tag{13.7}$$

Equation (13.7) is essentially equal to (13.4), but it cannot conveniently be applied to cases where $i \geq 2$.[38] Therefore, to derive an explicit expression from (13.7), we make the following simplifying assumptions. For the ith binding equilibrium, the binding rate of one A molecule to $n - i + 1$ vacant sites on a protein molecule is equal to the dissociation rate of one molecule from the protein molecule bound with i molecules of A, where the former is proportional to the number of vacant sites and the latter to the number of bound A molecules. We then have the following equation for the velocity of $[PA_i]$ change:

$$d[PA_i]/dt = (n - i + 1)\vec{k}[PA_{i-1}]C_f - i\overleftarrow{k}[PA_i] = 0 \tag{13.8}$$

or

$$K_i = [PA_i]/([PA_{i-1}]C_f) = (n - i + 1)k/i \qquad k = \vec{k}/\overleftarrow{k} \qquad (13.8')$$

where $\vec{k}$ is the intrinsic binding rate constant of an A molecule, $\overleftarrow{k}$ is its dissociation constant, and k is their ratio.

The following equality for K_1 results from the same analogy:

$$K_1 = [PA_1]/([P]C_f) = nk \qquad (13.9)$$

From (13.8') and (13.9) there results

$$K_i = (n - i + 1)K_1/ni \qquad (13.10)$$

The product in the denominator of (13.7) can then be simplified, with the help of (13.8'), to

$$\prod_{j=1}^{i} K_j = \{n!/[i!(n - i)!]\} \times k^i \qquad (13.11)$$

The r value then becomes

$$r = \sum_{i=1}^{n} i\{n!/[i!(n - i)!]\}(kC_f)^i \Big/ \left[1 + \sum_{i=1}^{n} \{n!/[i!(n - i)!]\}(kC_f)^i \right] \qquad (13.12)$$

By the binomial theorem, the denominator is equal to $(1 + kC_f)^n$, while the numerator can also be obtained in a closed form by mathematical manipulation[20]:

$$d(1 + kC_f)^n/d(kC_f) = n(1 + kC_f)^{n-1}$$

$$= \sum_{i=1}^{n} i\{n!/[i!(n - i)!]\}(kC_f)^{i-1} \qquad (13.13)$$

Equation (13.12) finally becomes

$$r = nkC_f(1 + kC_f)^{n-1}/(1 + kC_f)^n = nkC_f/(1 + kC_f) \qquad (13.14)$$

Equation (13.14) is identical to (13.2), which indicates that the assumptions listed above are fundamentally equal to those of the Langmuir equation and the Scatchard isotherm. Thus, the binding isotherms are idealized and do not take into consideration either cooperative binding or adsorption.

Because the binding of amphiphiles to proteins results from the hydrophobic and electrostatic interactions, cooperativity actually is very important, just as it is in micelle formation. In addition, the hydrophobic interaction between proteins and ionic surfactants results in a conformational change of the polymers even at remarkably low surfactant concentrations.[18] Such cooperative binding of low-molecular-weight amphiphiles to polymers has been well treated theoretically by Satake and Yang,[39] who adapted the Zimm–Bragg theory[40] for helix–coil transitions to the cooperative binding isotherm of ionic surfactants to polypeptides. The theory defines two parameters: the equilibrium constant s and an initiation factor σ. If the digit 0 represents an unbound polypeptide side chain and 1 a surfactant-bound polypeptide side chain, the state of the polypeptide chain can be schematically described by a sequence such as . . . 00001111100011100 . . . . The free energy change of the binding is so great per polypeptide residue, compared with that of its conformational change, that the conformational change need not be considered in this discussion. The binding between polymer residues and the surfactants can be written by either of the following two binding equilibria:

$$(00) + A \underset{}{\overset{K_0}{\rightleftharpoons}} (01) \tag{13.15}$$

$$(10) + A \underset{}{\overset{K_0 u}{\rightleftharpoons}} (11) \tag{13.16}$$

where K_0 is the binding constant of the surfactant molecule to a site with two unoccupied nearest neighbors, and $K_0 u$ is the binding constant of a site that follows one or more occupied sites. Here, u can be described in terms of the interaction energies E of the neighboring groups:

$$u = \exp[(2E_{01} - E_{11} - E_{00})/kT] \tag{13.17}$$

If the average electrical potential at the polymer surface is ψ_0, K_0 in (13.15) and (13.16) can be written as

$$K = K_0 \exp(-e\psi_0/kT) \tag{13.18}$$

Thus, denoting the concentration by brackets, we may write

$$\sigma s = [01]/[00] = KC_f \tag{13.19}$$

and

$$s = [11]/[10] = KuC_f \tag{13.20}$$

with

$$\sigma = [01][10]/[00][11] = 1/u \tag{13.21}$$

According to the Zimm–Bragg theory,[40] the degree of binding $x = C_b/C_p$, where C_p is the polypeptide residue concentration, becomes

$$x = d \ln \lambda_0/d \ln s \tag{13.22}$$

where λ_0 represents the larger of the two eigenvalues for the statistical weight matrix, and is given by

$$\lambda_0 = \{1 + s + [(1 - s)^2 + 4\sigma s]^{1/2}\}/2 \tag{13.23}$$

It follows, therefore, from (13.19), (13.20), (13.22), and (13.23), that

$$2x - 1 = (s - 1)/[(1 - s)^2 + 4\sigma s]^{1/2}$$

$$= (KuC_f - 1)/[(1 - KuC_f)^2 + 4KC_f]^{1/2} \tag{13.24}$$

In the above equations, K is assumed constant because ψ_0 remains virtually constant until x becomes large. Equation (13.24) is equivalent to the theoretical equation by Schmitz and Schurr.[41] The free surfactant concentration at $x = 0.5$ in (13.24) satisfies $Ku(C_f)_{x=0.5} = 1$, and (13.24) is rewritten as

$$2x - 1 = (y - 1)/[(1 - y)^2 + 4y/u]^{1/2} \tag{13.25}$$

where $y = C_f/(C_f)_{x=0.5}$. By differentiating x in (13.24) with respect to C_f and substituting $(C_f)_{x=0.5}$ by $1/Ku$, we have

$$(dx/d \ln C_f)_{x=0.5} = u^{1/2}/4 \tag{13.26}$$

Thus, the parameter value of u can be determined by the observed binding isotherm onto polymers. In addition, the number of bound surfactant ion clusters (Z) is given by $d \ln \lambda_0^n/d \ln \sigma$.[40] Hence, the average cluster size of bound surfactant ions $(\bar{m})$ and unoccupied binding sites $(\bar{p})$ can be derived from (13.23)

$$\bar{m} = nx/Z = 2x(u-1)/\{[4x(1-x)(u-1)+1]^{1/2}-1\} \quad (13.27)$$

and

$$\bar{p} = (1-x)\bar{m}/x \quad (13.28)$$

The above theoretical treatment is also equivalent to those of Marcus[42] and Lifson,[43] and has been employed for surfactant binding to polyelectrolytes.[44,45]

13.3. Binding of Amphiphiles to Synthetic Polymers

The finding that mixed solutions of polymers and amphiphiles often exhibit solution properties much different from those of each individual solution strongly suggests that the formation of a polymer–amphiphile complex results from interaction between polymers and amphiphiles. For example, the aqueous solubility of sodium hexadecyl sulfate (Krafft temperature range approximately 30°C) is less than $5 \times 10^{-4}\,\mathrm{mol \cdot dm^{-3}}$ at 25°C. However, it increases with the addition of polymers that form a water-soluble complex (Fig. 13.1).[6] The increase depends on the kind of polymers, their molecular weight and concentrations, and temperature. Amylose molecules precipitate with sodium dodecyl sulfates as a clathrate compound.[46,47] These compounds are formed mainly by hydrophobic interaction between polymers and surfactants, because nonionic polymers are also able to form complexes and because only surfactants with alkyl chains longer than a certain carbon number form complexes. Specifically, only anionic surfactants with alkyl chains longer than C_8, and only cationic surfactants with alkyl chains longer than C_{14}, form complexes with polyvinyl alcohol, polyvinyl pyrrolidone, and polyethylene glycol. Anionic surfactants also bind much more extensively to polymers than do cationic surfactants.

The concentration range in which surfactants begin to bind to polymers is very narrow, much as for micelle formation, which strongly suggests a cooperative binding mechanism. The relation between the logarithm of the concentration and surfactant carbon number is linear,[5,48,49] and the free energy of transferring a methylene group from the aqueous bulk to the complex is also equivalent to that of micelle formation. The solubilization of sparingly soluble materials by the complex yields plots of maximum additive concentration (MAC) versus surfactant concentration similar to those given by pure micelles. The MAC generally increases with polymer concentration at a fixed surfactant concentration.[50,51] The function of the

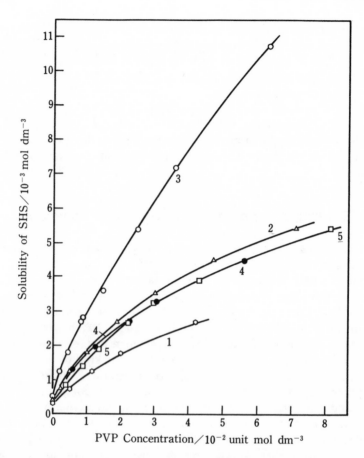

Figure 13.1. Solubility of sodium hexadecyl sulfate (SHS) in aqueous solutions of polyvinyl-pyrrolidone of difference molecular weight (MW) as a function of polyvinylpyrrolidone (PVP) concentration.[6] 1, MW = 28,000 at 20°C; 2, MW = 28,000 at 25°C; 3, MW = 28,000 at 30°C; 4, MW = 13,000 at 25°C; 5, MW = 220,000 at 25°C. (Reproduced with permission of Steinkopf-Verlag Darmstadt.)

polymer in such a complex is to make the surface of the complex more hydrophobic and thus to promote the interaction between the complex and the solubilizate. Kinetic studies of polymer–surfactant complex formation support the idea that the complex consists of a micelle wrapped in a polymer chain rather than a polymer saturated by linear adsorption of surfactant molecules.[52] The adsorbed polymer localizes in the micelle palisade layer or at the micellar surface, leading to an increase in the micelle ionization of ionic surfactants,[53-55] and the micellar aggregation number of the complexes decreases with increasing polymer concentration.[55,56]

Many reports have appeared regarding complex formation for a variety of systems, such as nonionic and anionic surfactants with sodium carboxy-methylcellulose,[57] anionic and cationic surfactants with nonionic polymers,[58] SDS with methylcellulose and polyvinyl alcohol,[59] and SDS with gelatin.[60] Reports have also dealt with such topics as the effect of ionic surfactant counterions on complex formation,[61] the dependence of folding and unfolding of polymers on the extent of binding,[62] and the photochemistry of the complex.[63]

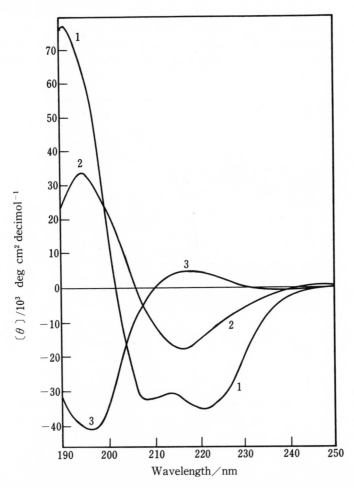

Figure 13.2. CD spectra of poly(L-lysine).[71] 1, α-helix at pH = 11.1; 2, β-structure obtained by heat treatment at 52°C for 15 in at pH = 11.1; 3, random coil at pH = 5.7. (Reproduced with permission of the American Chemical Society.)

Synthetic polypeptides, although not true proteins, have been used as a model for studying the effects of amphiphiles on the secondary structure of proteins. Poly(L-lysine) (PLL) and poly(L-glutamic acid) (PLG) are typical polyelectrolytes at neutral pH because the head groups of their side chains are completely dissociated ($pK_a = 10.04$[64] for PLL and $pK_a = 4.45$[65] for PLG). Polypeptide chains experiencing electrostatic repulsion between the ionized head groups should not adopt an ordered conformation such as an α-helix or β-structure, but instead should display a random coil conformation. The β-structure can be induced by a heat treatment, and is strongly dependent on polypeptide concentration.[66] However, reducing the degree of ionization leads to the helix conformation, where the coil-to-helix change is reversible and is easier for longer hydrophobic side chains.[64,67] A conformational change has also been observed to be induced by binding

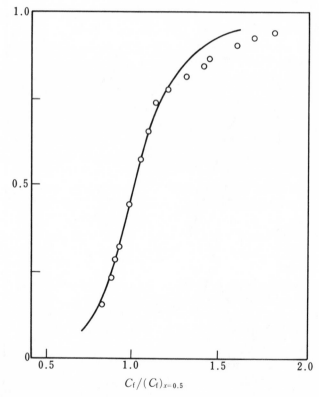

$$C_f/(C_f)_{x=0.5}$$

Figure 13.3. Calculated and observed binding isotherm of sodium decyl sulfate for poly(L-ornithine) at 25°C.[39] O, observed; ———, from Eq. (13.25) with $u = 77$. (Reproduced with permission of Wiley & Sons.)

of surfactants to these polypeptides. These conformational changes can be examined by spectroscopic methods such as UV, circular dichroism (CD), and optical rotatory dispersion (ORF), and the spectrum has been determined for each conformation.[68-73] Figure 13.2 shows typical examples of CD spectra for the three conformations of PLL.[71]

Polypeptides are very suitable for examining the cooperative binding of surfactants to polymers. Satake and Yang determined the binding isotherms of sodium decyl sulfate onto several polypeptides and applied their theory of cooperative binding to the isotherms (Fig. 13.3),[39] where the hydrophobic interaction between bound surfactant ions was related to the stability of the polypeptide conformations.

13.4. Binding of Amphiphiles to Proteins

The existence of hydrophobic amino acid side chains on protein molecules gives rise to an interaction between proteins and hydrophobic solutes, such as synthetic amphiphiles and biological materials *in vivo*. Because the side chains of several amino acids carry an amino or carboxyl group, the overall electrical charge of a protein molecule depends on its configuration and on the solution pH. Variety of the functions of proteins in living membranes and in the bloodstream originates from their association with amphiphiles due to the above interactions, which are of course pH and temperature dependent. The binding of solutes to macromolecules depends more on electrostatic than on hydrophobic interaction, as would be expected. For example, the binding sites of native serum albumin for anionic amphiphiles are quite specific, whereas binding sites for neutral and cationic amphiphiles are rarely observed.[74] A combination of a hydrophobic region and an ionic region might become the binding site for an amphiphile absorbate.[19] It is also possible that the alkyl chain length of the absorbates may or may not have an effect on the magnitude of the binding constant.

The interaction between proteins and amphiphiles has been extensively examined from the viewpoint of adsorption or binding, where the Langmuir type of adsorption has most often been suggested. However, many adsorption isotherms do not exhibit saturation but, rather, increase steadily with the adsorbate concentration. This finding has two possible explanations: either the adsorbate concentration cannot achieve the saturating adsorption conditions, or nonspecific adsorption sites are present in addition to the specific sites. In the latter case, adsorption to the specific sites is stong and leads to saturation (curve C in Figure 13.4), whereas adsorption to the nonspecific sites is weak and increases linearly with the concentration in

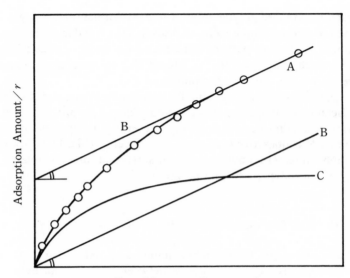

Free Monomer Concentration

Figure 13.4. Change of adsorption amount with concentration of free monomer (C_f).[99] A, total adsorption; B, due to nonspecific binding; C, due to specific binding.

the bulk solution (curve B). The total amount of adsorption increases with the adsorbate concentration in the bulk solution (curve A), and becomes a sum of the above two:

$$r = \sum_i [n_i K_i C_f / (1 + K_i C_f)] + a C_f \qquad (13.29)$$

where a is the coefficient for the nonspecific adsorption.

The adsorption isotherm of methyl orange onto albumin resembles curve C,[75] whereas 8-anilino-1-naphthalene sulfonate adsorbes to albumin in accordance with curve A.[76] Adsorption of more than a certain critical amount of synthetic surfactants causes a denaturation of the protein, resulting in appearance of further binding sites.[77-79] Therefore, the adsorption isotherm is no longer a single sigmoid curve, and adsorption is accompanied by changes in protein conformation and by other adsorption mechanisms, which can be easily detected by optical methods. Denaturation of proteins involves unfolding of the protein. Disulfide bridges and –SH groups inside the protein move to the surface and undergo an internal or external exchange reaction between the bonds and the groups. The minimum amphiphile concentration to bring about this exchange reaction becomes less with increasing alkyl chain length.[80]

If the binding process is cooperative, a steep increase in the average number of molecules bound per protein molecule (r) will be observed within a narrow concentration range of monomeric adsorbate (just as for micelle formation by surfactants). To determine the mechanisms of adsorption and binding, systematic adsorption experiments with homologous adsorbates differing only in alkyl chain length are indispensable. Interactions involving hydrophilic head groups can be assumed to be identical for such an adsorbate series, and the effect of alkyl chain length can be taken to reflect only the hydrophobic part of the interaction. When hydrophobic interaction plays an important role in binding, the adsorption amount of the adsorbate with longer alkyl chain increases.[81] However, the standard free energy change of binding starts to deviate from linearity above $n = 8$ when the former is plotted against the number n of carbon atoms in the adsorbate.[82] The slope below the number corresponds to the free energy change of transfer per methylene group from the aqueous environment to the nonpolar solvent. This deviation suggests that the "excess" methylene groups (more than 8) are located partly in the aqueous environment.

The above model is not always applicable.[35] In particular, when the adsorption amount becomes high enough to stimulate cooperative binding and to solubilize the protein from an organic tissue, surfactants will micellize

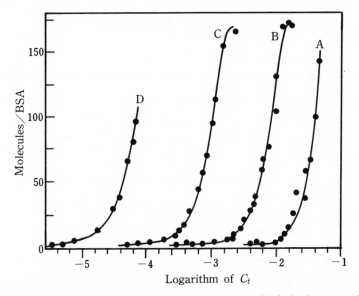

Figure 13.5. Binding isotherms of alkyltrimethylammonium bromide for bovine serum albumin (BSA) at pH = 6.9, ionic strength = 0.1, and 25°C.[35] A, decyl-; B, dodecyl-; C, tetradecyl-; D, hexadecyl-. (Reproduced with permission of the Chemical Society of Japan.)

on the protein surface with their ionic head groups contacting the aqueous environment and their hydrophobic tails contacting the protein.[83-87] Figure 13.5 shows the adsorption isotherms of the cooperative binding of alkyltrimethylammonium bromides onto bovine serum albumin.[35] The adsorption at concentrations below the steep increase represents binding to independent sites, whereas the steep increase of adsorption corresponds to cooperative binding with a positive entropy change.

A change in the charge of the surfactant ion from negative to positive not only affects the extent of binding but also can result in alterations in protein structure due to changes in surfactant binding.[80,88] The formation of complexes between anionic surfactants and proteins is well established.[89-92] Reports are also available on adsorption of inorganic ions to human serum albumin,[93] solubilization by protein and surfactant complex,[94] denaturation of bovine serum albumin by stepwise binding SDS,[95] interaction between anesthetic agents and proteins,[96] binding of fatty acids to albumin from the viewpoint of protein denaturation,[97] and binding of bile acids to lysozyme.[98]

References

1. H. B. Bull and H. Neurath, *J. Biol. Chem.* 118, 163 (1937).
2. N. Sata and S. Saito, *Kolloid Z.* 128, 154 (1952).
3. S. Saito, *Kolloid Z.* 137, 98 (1954).
4. M. N. Jones, *J. Colloid Interface Sci.* 23, 36 (1967).
5. M. J. Schwuger and H. Lange, 5th International Congress on Surface Active Substances, Barcelona, 1968, Vol. II, p. 955 (1969).
6. S. Saito, *Kolloid Z. Z. Polym.* 215, 16 (1967).
7. S. Saito, *Kolloid Z.* 133, 12 (1953).
8. M. Nakagaki and Y. Ninomiya, *Bull. Chem. Soc. Jpn.* 37, 817 (1964).
9. A. Arai, F. Shigehiro, and I. Maruta, *Kogyo Kagaku Zasshi* 68, 1090 (1965).
10. W. Fong and W. H. Ward, *Text. Res. J.* 24, 881 (1954).
11. T. Isemura and A. Imanishi, *J. Polym. Sci.* 33, 337 (1958).
12. C. Botre, F. DeMartiis, and M. Solinas, *J. Phys. Chem.* 68, 3624 (1964).
13. S. Saito, *Kolloid Z.* 154, 19 (1957).
14. F. Tokiwa and N. Moriyama, *J. Colloid Interface Sci.* 30, 338 (1969).
15. F. W. Putnam, *Adv. Protein Chem.* 4, 79 (1948).
16. J. F. Foster, in: *The Plasma Proteins* (F. W. Putnam, ed.) Vol. 1, p. 179, Academic Press, New York (1960).
17. M. Joly, *A Physico-Chemical Approach to the Denaturation of Proteins*, p. 30, Academic Press, New York (1965).
18. J. Steinhardt and J. A. Reynolds, *Multiple Equilibria in Proteins*, Academic Press, New York (1969).
19. C. Tanford, *The Hydrophobic Effect: Formation of Micelles and Biological Membranes*, p. 126, Wiley, New York (1973).
20. M. N. Jones, *Biological Interfaces*, p. 101, Elsevier, Amsterdam (1975).

21. M. L. Anson, *J. Gen. Physiol. 23*, 239 (1939).
22. A. L. Shapiro, E. Vinuela, and J. V. Maizel, Jr., *Biochem. Biophys. Res. Commun. 28*, 815 (1967).
23. K. Weber and M. Osborn, *J. Biol. Chem. 244*, 4406 (1969).
24. S. A. Rosenberg and G. Guidotti, *J. Biol. Chem. 244*, 5118 (1969).
25. G. Guidotti, *Ann. Rev. Biochem. 41*, 731 (1972).
26. J. R. Huizenger, P. F. Grieger, and F. T. Wall, *J. Am. Chem. Soc. 72*, 2636 (1950).
27. M. J. Pallansch and D. R. Briggs, *J. Am. Chem. Soc. 76*, 1396 (1954).
28. G. Strauss and U. P. Strauss, *J. Phys. Chem. 62*, 1321 (1958).
29. A. Ray, J. Reynolds, H. Polet, and J. Steinhardt, *Biochemistry, 5*, 2606 (1966).
30. J. Reynolds, H. Herbert, H. Polet, and J. Steinhardt, *Biochemistry 6*, 937 (1967).
31. A. V. Few. R. H. Ottewill, and H. C. Parreira, *Biochim. Biophys. Acta 18*, 136 (1955).
32. S. Kaneshina, M. Tanaka, T. Kondo, T. Mizuno, and K. Aoki, *Bull. Chem. Soc. Jpn. 46*, 2735 (1973).
33. Y. Nozaki, J. A. Reynolds, and C. Tanford, *J. Biol. Chem. 249*, 4452 (1974).
34. M. N. Jones, H. A. Skinner, and E. Tipping, *Biochem. J. 147*, 229 (1975).
35. K. Hiramatsu, C. Ueda, K. Iwata, K. Arikawa, and K. Aoki, *Bull. Chem. Soc. Jpn. 50*, 368 (1977).
36. G. Scatchard, *Ann. N. Y. Acad. Sci. 51*, 660 (1949).
37. I. Langmuir, *J. Am. Chem. Soc. 40*, 1361 (1918).
38. I. M. Klotz and D. L. Hunston, *Biochemistry 10*, 3065 (1971).
39. I. Satake and J. T. Yang, *Biopolymers 15*, 2263 (1976).
40. B. H. Zimm and J. K. Bragg, *J. Chem. Phys. 31*, 526 (1959).
41. K. S. Schmitz and J. M. Schurr, *Biopolymers 9*, 697 (1970).
42. R. A. Marcus, *J. Phys. Chem. 58*, 621 (1954).
43. S. Lifson, *J. Chem. Phys. 26*, 727 (1957).
44. K. Hayakawa and J. C. T. Kwak, *J. Phys. Chem. 86*, 3866 (1982).
45. K. Hayakawa and J. C. T. Kwak, *J. Phys. Chem. 87*, 506 (1983).
46. T. Takagi and T. Isemura, *Bull. Chem. Soc. Jpn. 33*, 437 (1960).
47. E. M. Osman, S. J. Leith, and M. Fles, *Cereal Chem. 38*, 449 (1961).
48. H. Arai, M. Murata, and K. Shinoda, *Bull. Chem. Soc. Jpn. 37*, 223 (1971).
49. K. Shirahama and N. Ide, *J. Colloid Interface Sci. 54*, 450 (1976).
50. S. Saito, *J. Colloid Interface Sci. 24*, 227 (1967).
51. F. Tokiwa and K. Tsujii, *Bull. Chem. Soc. Jpn. 46*, 2684 (1973).
52. C. Tondre, *J. Phys. Chem. 89*, 5101 (1985).
53. Y. Moroi, H. Akisada, M. Saito, and R. Matuura, *J. Colloid Interface Sci. 61*, 233 (1977).
54. H. Akisada, Y. Kuroki, K. Koga, Y. Moroi, and R. Matuura, *Mem Fac. Sci. Kyushu Univ. Ser. C 10*, 189 (1978).
55. R. Zana, P. Lianos, and J. Lang, *J. Phys. Chem. 89*, 41 (1985).
56. E. A. Lissi and A. Abuin, *J. Colloid Interface Sci. 105*, 1 (1985).
57. M. J. Schwuger and H. Lange, *Tenside 5*, 257 (1968).
58. M. J. Schwuger, *J. Colloid Interface Sci. 43*, 491 (1973).
59. K. E. Lewis and C. P. Robinson, *J. Colloid Interface Sci. 32*, 539 (1970).
60. W. J. Knox, Jr., and T. O. Parshall, *J. Colloid Interface Sci. 33*, 16 (1970).
61. S. Saito, T. Taniguchi, and K. Kitamura, *J. Colloid Interface Sci. 37*, 154 (1971).
62. S. Saito and T. Taniguchi, *J. Colloid Interface Sci. 44*, 114 (1973).
63. K. Hayakawa, J. Ohta, T. Maeda, I. Satake, and J. C. T. Kwak, *Langmuir 3*, 377 (1987).
64. M. J. Grourke and J. H. Gibbs, *Biopolymers 10*, 795 (1971).
65. M. Nagasawa and A. Holtzer, *J. Am. Chem. Soc. 86*, 538 (1964).
66. S. Y. C. Wooley and G. Holzwarth, *Biochemistry 9*, 3604 (1970).
67. S. R. Chaudhuri and J. T. Yang, *Biochemistry 7*, 1379 (1968).

68. B. Davidson, N. Tooney, and G. D. Fasman, *Biochem. Biophys. Res. Commun.* 23, 156 (1966).
69. R. Townend, T. F. Kumosinski, S. N. Timasheff, G. D. Fasman, and B. Davidson, *Biochem. Biophys. Res. Commun.* 23, 163 (1966).
70. N. Greenfield, B. Davidson, and G. D. Fasman, *Biochemistry 6*, 1630 (1967).
71. N. Greenfield and G. D. Fasman, *Biochemistry 8*, 4108 (1969).
72. G. Holzwarth and P. Doty, *J. Am. Chem. Soc. 87*, 218 (1965).
73. J. T. Yang, in: *Conformational Biopolymers* (G. N. Ramachandran, ed.) Vol. I, p. 157, Academic Press, New York (1967).
74. J. Reynolds, S. Herbert, and J. Steinhardt, *Biochemistry 7*, 1357 (1968).
75. K. Shikama, *J. Biochem. 64*, 55 (1968).
76. E. C. Santos and A. A. Spector, *Biochemistry 11*, 2299 (1972).
77. R. V. Decker and J. F. Foster, *Biochemistry 5*, 1242 (1966).
78. R. D. Hagenmaier and J. F. Foster, *Biochemistry 10*, 637 (1971).
79. O. Laurie and J. Oakes, *J. Chem. Soc. Faraday Trans. 1 72*, 1324 (1976).
80. K. Aoki and K. Hiramatsu, *Anal. Biochem. 60*, 213 (1974).
81. D. S. Goodman, *J. Am. Chem. Soc. 80*, 3892 (1958).
82. C. Tanford, *J. Mol. Biol. 67*, 59 (1972).
83. J. Steinhardt, N. Stocker, D. Carroll, and K. S. Birdi, *Biochemistry 13*, 4461 (1974).
84. K. Takeda, *Bull. Chem. Soc. Jpn. 55*, 2547 (1982).
85. D. M. Bloor and E. Wyn-Jones, *J. Chem. Soc. Faraday Trans. 2 78*, 657 (1982).
86. J. Oakes, *J. Chem. Soc. Faraday Trans. 1 70*, 2200 (1974).
87. Y. Inoue, S. Sase, R. Chujo, S. nagaoka, and M. Sogami, *Biopolymers 18*, 373 (1979).
88. K. Hiramatsu, *Biochim. Biophys. Acta 490*, 209 (1977).
89. F. W. Putnam and H. Neurath, *J. Am. Chem. Soc. 66*, 692 (1944).
90. J. T. Yang and J. F. Foster, *J. Am. Chem. Soc. 75*, 5560 (1953).
91. T. Isemura, F. Tokiwa, and S. Ikeda, *Bull. Chem. Soc. Jpn. 28*, 555 (1955); *35*, 240 (1962).
92. W. J. Knox, Jr., and J. F. Wright, *J. Colloid Sci. 20*, 177 (1965).
93. G. Scatchard, I. H. Scheinberg, and S. H. Armstrong, Jr., *J. Am. Chem. Soc. 72*, 535, 540 (1950).
94. M. J. Schwuger, *Kolloid Z. Z. Polym. 233*, 898 (1969).
95. K. Takeda, M. Miura, and T. Takagi, *J. Colloid Interface Sci. 82*, 38 (1981).
96. M. Abu-Hamidiyyah, *Langmuir 2*, 310 (1986).
97. K. Aoki, N. Hayakawa, K. Noda, H. Terada, and K. Hiramatsu, *Colloid Polym. Sci. 261*, 359 (1983).
98. Y. Murata, M. Okawauchi, H. Kawamura, G. Sugihara, and M. Tanaka, in: *Surfactants in Solution* (K. L. Mittal and P. Bothorel, eds.), Vol. 5, p. 861, Plenum Press, New York (1987).
99. Z. Taira and H. Terada, *Biochem. Pharmacol. 34*, 1999 (1985).

Index